Math Field Trip:
A Grade-Raising Dictionary for Students Ages 9–12

Math Field Trip

A Grade-Raising Dictionary for Students Ages 9–12

By Jeanine Le Ny
with Meg Moyer, Math Consultant

New York • Chicago

Editorial Director: Jennifer Farthing
Editor: Cynthia Ierardo
Production Editor: Dave Shaw
Cover Designer: Carly Schnur

Published by Kaplan Publishing, a division of Kaplan, Inc.
888 Seventh Ave.
New York, NY 10106

Printed in the United States of America

December 2006
07 08 09 10 9 8 7 6 5 4 3 2 1

Library of Congress Cataloging-in-Publication Data

Le Ny, Jeanine.
Math field trip : a grade-raising math dictionary for students ages 9-12 : a fun and exciting way to learn 250 math terms and concepts / by Jeanine Le Ny ; with Meg Moyer, Math Consultant.—1st ed.
p. cm.
ISBN-13: 978-1-4195-9149-5
ISBN-10: 1-4195-9149-5
1. Mathematics—Dictionaries, Juvenile. I. Moyer, Meg. II. Title.
QA5.L4 2006
510.3--dc22 2006030584

Kaplan Publishing books are available at special quantity discounts to use for sales promotions, employee premiums, or educational purposes. Please call our Special Sales Department to order or for more information at 800-621-9621, ext. 4444, e-mail kaplanpubsales@kaplan.com, or write to Kaplan Publishing, 30 South Wacker Drive, Suite 2500, Chicago, IL 60606-7481.

Have you ever felt like learning math is a bore? Sometimes we just learn math because we *have* to. We memorize it for a test, and then quickly forget it when it's over. Well, this book teaches math in a different way! *The Math Field Trip* takes you on an adventure with 7 friends who are visiting the science museum.

When you have an interesting group of kids on a tour of the museum, fun, trouble, and mishaps are bound to happen!

Read about **Johnny,** the fun-loving guy who sometimes gets himself in trouble; his friend **Chris,** the sneaky kid who is trying to ditch his museum buddy; Anna, the brainy girl who is eager to learn about everything; **Lester,** the nerdy student who likes Anna and loves math; **Wendy, Gabby, and Sonia,** the three best friends and most popular girls in the class; and **Mr. Stickler,** the children's beloved teacher with a big sweet tooth!

With *Math Field Trip,* Kaplan offers you:

- 250 important math terms and concepts
- An entertaining story and cast of characters
- Sample sentences to help you apply the math terms to everyday situations

So join these characters on their adventures in visiting the different exhibits at the museum. Learn what they discover about science and each other, and learn some excellent math skills along the way!

ENJOY THE FIELD TRIP!

A

Absolute value ***The distance between a number and zero on the number line; the absolute value of any number is positive because distance is always positive.***

Johnny was so bored on the bus ride to the science museum, he used a number line to chart the amount of yellow cars he spotted and came up with an **absolute value** of 2.

|–6 |=

Acute angle ***An angle that measures more than 0° and less than 90°.***

Anna made sure that her novel was opened at an **acute angle** so that the boy sitting next to her on the bus could not see what she was reading.

Which of the following angles is an acute angle?

(A) 180° (C) 50° (B) 130° (D) 90°

Acute triangle *A triangle with three acute angles.*
Best friends Wendy, Gabby, and Sonia gave one another their special secret greeting: forming an **acute triangle** with their pinkies.

Which of the following represents the measures of the angles of an acute triangle?
(A) 30°, 60°, and 90° (C) 40°, 60°, and 80°
(B) 25°, 45°, and 110° (D) 22°, 58°, and 100°

Add (+) *To join 2 or more items or groups.*
Chris **added** up the change in his pocket, hoping he'd have enough money to buy a souvenir at the museum.

Add 27 and 41.

Adjacent angles *Angles that share a common vertex and side and have no interior points in common.*
Lester made sure that he sat **adjacent** to the toilet, just in case he had to barf on the bus.

Which of the following angles are adjacent?

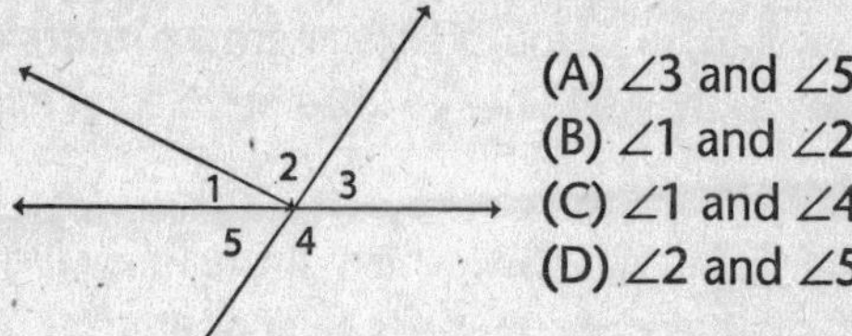

(A) ∠3 and ∠5
(B) ∠1 and ∠2
(C) ∠1 and ∠4
(D) ∠2 and ∠5

Alternate exterior angles *Two angles that lie on the exterior of 2 different lines and on different sides of the transversal.*
Wendy and Gabby sat back-to-back in their seats, forming **alternate exterior angles.**

Which of the following pairs of angles are alternate exterior angles?

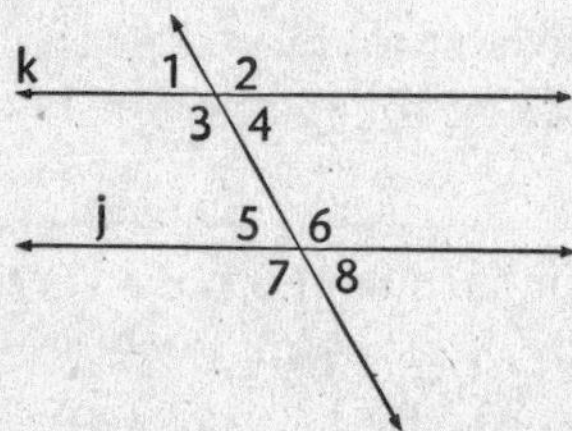

(A) ∠1 and ∠2 (C) ∠3 and ∠6
(B) ∠1 and ∠5 (D) ∠1 and ∠8

Alternate interior angles *Two angles that lie on the interior of 2 different lines and on different sides of the transversal.*

Gabby and Sonia looked like **alternate interior angles** as they sat knee-to-knee huddled in gossip.

In the diagram below, which of the following pairs of angles are alternate interior angles?

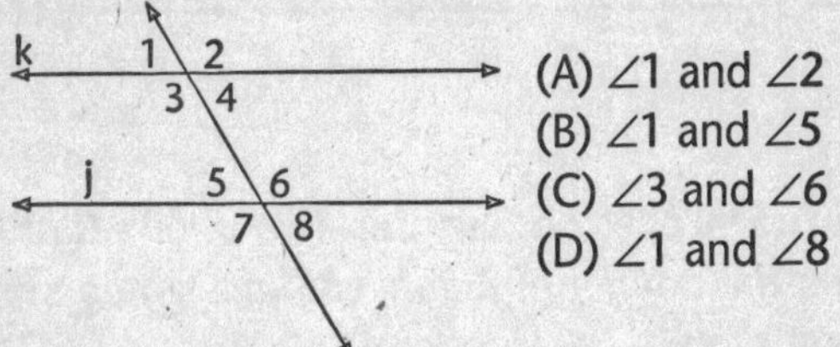

(A) ∠1 and ∠2
(B) ∠1 and ∠5
(C) ∠3 and ∠6
(D) ∠1 and ∠8

Angle *Two rays joined at a common endpoint.*

Instead of waving to his students, Mr. Stickler raised a hand and formed an **angle** by separating his index and middle fingers, which is commonly known as the peace sign.

Which of the following is *not* a way to name the angle shown below?

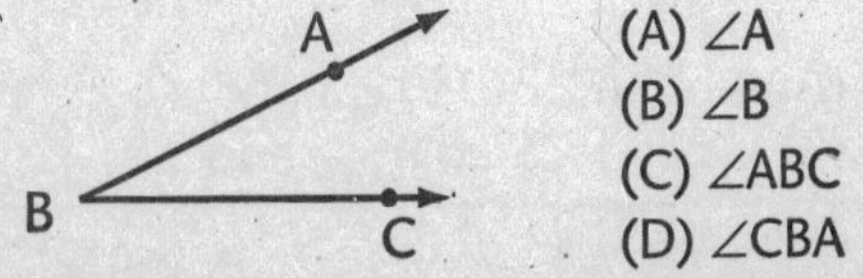

(A) ∠A
(B) ∠B
(C) ∠ABC
(D) ∠CBA

Area ***The measurement of the amount of space enclosed in a two-dimensional figure.***

Lester wondered how much carpet it took to fill the **area** of the bus aisle.

What is the area of a rectangle whose base is 10 and height is 3?

Associative property ***Changing the grouping does not change the sum or product.***

Associative Property of Addition:

$$(a + b) + c = a + (b + c)$$

Associative Property of Multiplication:

$$(ab)c = a(bc)$$

The **associative property** states that it doesn't matter if Gabby and Wendy share a seat on the bus and Sonia sits in the seat next to them *or* if Gabby sits alone and Wendy and Sonia share a seat. They'll all still be best friends.

Which of the following is an example of the associative property of multiplication?

(A) $(6 \times 2) \times 3 = 6 \times (2 \times 3)$
(B) $(6 + 2) + 3 = 6 + (2 + 3)$
(C) $3 \times 5 = 5 \times 3$
(D) $5(4 + 2) = 5 \times 4 + 5 \times 2$

Average ***The sum of the values divided by the number of values; also called the mean.***
Johnny's parents told him that that he had to raise his grade point **average** in order to go on the class trip.

A student received scores of 90, 86, 81, and 95 on his first four tests. What is his average test score?

Axes ***Two perpendicular number lines in the coordinate system; the vertical line is the y-axis and the horizontal line is the x-axis.***
Mr. Stickler asked the class to form two straight y-**axes** in front of the museum steps.

Is the point (4, 0) on the *x*-axis or the *y*-axis?

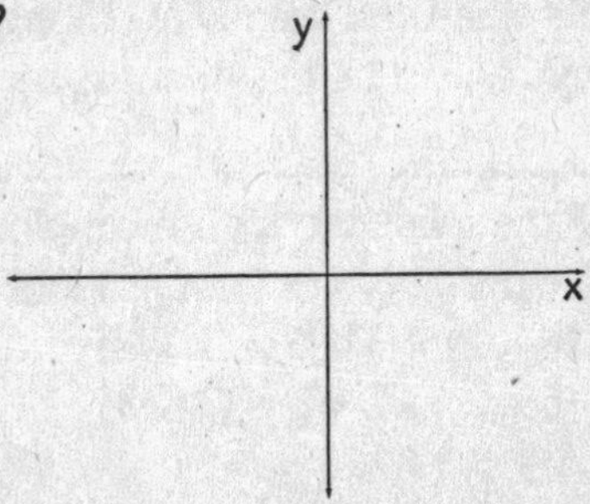

y

x

Bar graph ***A graph that uses vertical or horizontal bars to compare quantities.***

Unfortunately for Johnny, Mr. Stickler had a **bar graph** that documented the boy's monthly detention rates. The teacher would not be letting Johnny out of his sight.

According to the bar graph below, how many more students like baseball more than basketball?

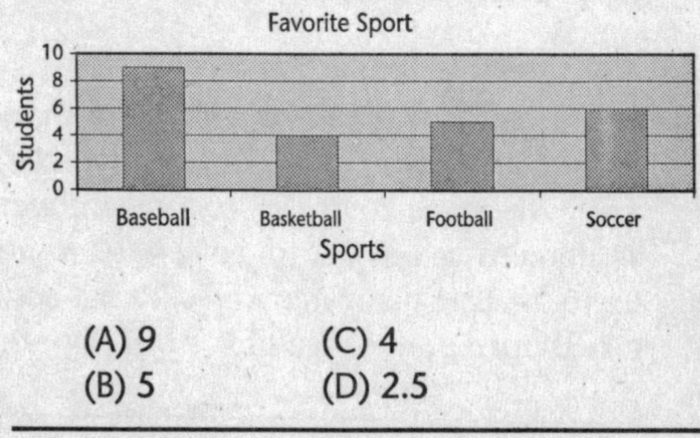

(A) 9
(B) 5
(C) 4
(D) 2.5

Base of a polygon *Any side may be called the base; the base must be perpendicular to the height.*

When the museum curator chided Chris for removing the cover of the square-shaped fish tank in the center of the lobby, Chris apologized for messing with the **base of a polygon.**

In the diagram below, which of the following sides is not a base of ΔEFG?

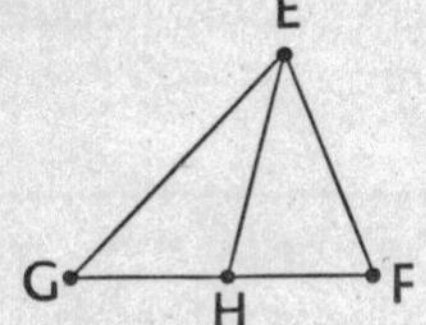

(A) $\overline{EH}$
(B) $\overline{EF}$
(C) $\overline{FG}$
(D) $\overline{EG}$

Base of a solid figure *In a pyramid or cone, the face on which the figure rests; in a cylinder or prism, there are 2 parallel, congruent bases.*

Wanting to be the first to see it, Anna rushed up to the Egyptian pyramid replica but accidentally tripped at the **base of the solid figure.**

Does a cylinder always rest on one of its bases?

Base of an exponent *A number (or quantity) which is raised to a power; 4 is the base in 4^3.* To some, Wendy might be considered a **base of an exponent** because her best friends, Sonia and Gabby, copy everything she says and does.

Compute 2^3.

Base ten *A number system containing 10 digits {0, 1, 2, 3, 4, 5, 6, 7, 8, and 9} in which the value of a digit is found by multiplying a digit by its place value.*

Mr. Stickler was thrilled when the museum's ticket agent said that the class's tickets cost "three". . . until he realized that the agent was using a **base ten** number system and really meant three hundred.

For example, in the number 32,600, the 3 has a value of thirty thousand and the 2 has a value of two thousand.

In the number 4,670, what is the value of the 6?

Binomial *A polynomial with 2 terms.*

When Mr. Stickler told the students to pick a museum buddy, Lester noted that they'd be kind of like **binomials.** Then Wendy noted that Lester was weird.

Which of the following expressions is a binomial?

(A) $4x$ (C) $x + 5$

(B) -7 (D) $x + y + z$

Box-and-whisker plot *A diagram that shows how a set of data is spread out; a box is drawn around the median and the upper and lower quartiles and "whiskers" extend out to the lowest and highest values.*

Mr. Stickler drew a **box-and-whisker plot** on his guide map to show the students which exhibits he'd like them to view.

In the box-and-whisker plot below, which number represents the median of the set of data?

5 14 24 45 62

(A) 5 (B) 14 (C) 24 (D) 45

Capacity ***The maximum amount a container can hold.***

The students had to wait to enter the Egyptian temple replica because it had a **capacity** of only ten people.

A shipping company makes boxes in the sizes shown in the table. Which could you use to ship an 8 in. × 10 in. × 12 in. treasure chest?

	Length (in Inches)	Width (in inches)	Height (in inches)
Box A	9	11	12
Box B	8	8	8
Box C	10	10	10
Box D	7	9	11

Celsius ***A way to measure temperature in the metric system; 0° C is the freezing point of water; 100° C is the boiling point of water.***

Chris laughed when he saw a display of a caveman who was wearing nothing but a fur loincloth. What did the guy do when it was only 0° **Celsius** outside?

Use the formula $F = \frac{9}{5}C + 32$, where F is Fahrenheit and C is Celsius, convert 25° Celsius to Fahrenheit.

Central angle ***An angle whose vertex is at the center of a circle.***

Lester passed by the cafeteria and noticed a lone triangle of pizza sitting on a silver tray, which reminded him of a **central angle** . . . and that he had math homework due tomorrow.

A central angle is drawn in which of the following circles?

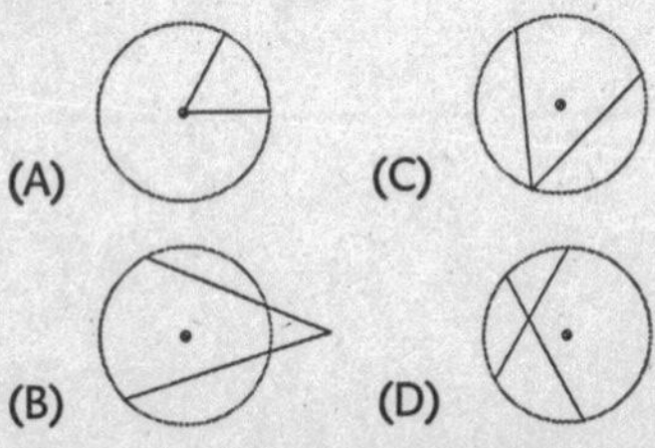

Chord ***A line segment whose endpoints are on the circle.***

Sonia marveled at an electricity display that had **chords** of lightening shooting around a glass globe.

In circle G, which segment is a chord?

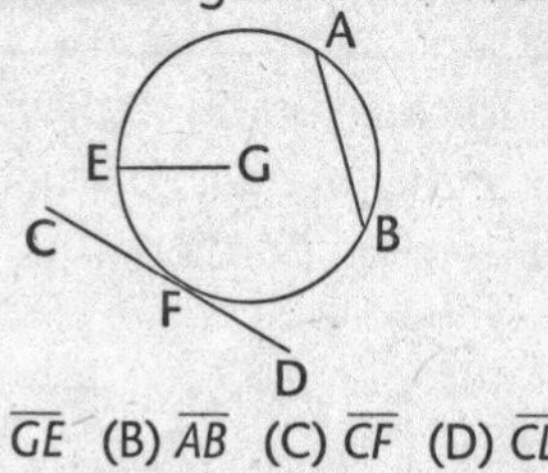

(A) $\overline{GE}$ (B) $\overline{AB}$ (C) $\overline{CF}$ (D) $\overline{CD}$

Circle ***A closed plane figure with all points an equal distance away from a fixed point, the center.***

Wendy drew a **circle** around the words "Experiment Gallery" on her museum brochure so she wouldn't forget to go there.

Which of the circles to the right has the longest diameter?

(A) (C)

(B) (D)

Circle graph ***A pie shaped graph that uses slices of different sizes to display data.***

If a scientist used a **circle graph** to record Lester's thoughts, 25% of the circle would be filled with pizza, 30% would be filled with math terms, and 45% would be filled with super-duper fudge chocolate chip brownies.

The following circle graph displays a college student's monthly budget. Estimate the % of her budget used on housing.

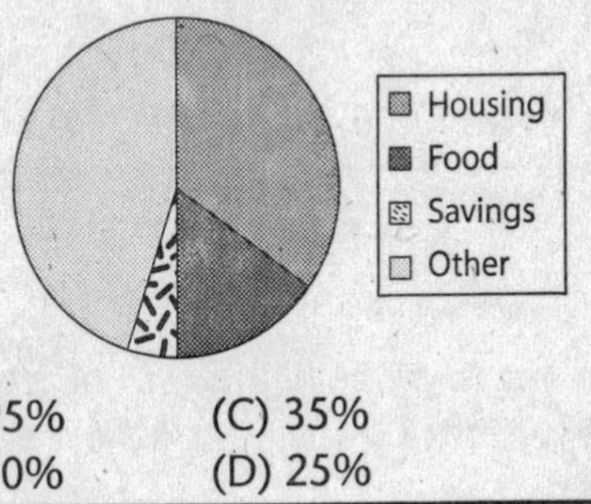

(A) 95% (C) 35%
(B) 50% (D) 25%

Circumference ***The distance around a circle; the formula for circumference is*** $C = \pi \times$ ***diameter*** $= \pi d$

$C = \pi d = 2\pi r.$

Anna sighed when she caught Lester staring at her because she knew he was trying to figure out the **circumference** of her blue eyes again.

Which of the carousels in the table below has the greatest circumference?

Carousel	Animals	Radius (yards)
Candy Carousel	30 horses, 5 giraffes, 2 pigs, 7 cows, 1 rooster	10
Blue Ribbon Carousel	18 horses, 1 lion, 3 elephants, 5 deer, 10 monkeys	12
Cascade Carousel	37 horses, 6 sheep, 1 hippo	14

Coefficient ***The number directly in front of the variable; it is connected to the variable by multiplication; 3 is the coefficient in the term $3a^4bc^2$; 1 is the coefficient in the term x^2.***
Gabby acted like a **coefficient** to her best friend Wendy—she was always right there in front of Wendy, multiplying her excitement in any situation.

Which of the following is the coefficient of the term $4x^2$?

(A) 4 (C) x
(B) 2 (D) x^2

Common denominator *The same number (or quantity) in the denominator of 2 or more fractions; when fractions have a common denominator, they can be combined using addition and/or subtraction.*

Wendy, Sonia, and Gabby's outfits could totally be mixed and matched because they all had a **common denominator:** a really hot pair of denim Capris.

Compute: $\frac{1}{2} + \frac{1}{5}$

Common factor *A whole number that is a factor of each number in a set of numbers; 5 is a common factor of 10, 15, and 20.*

Wendy is the **common factor** between Gabby and Sonia. Without Wendy, the two girls would probably have nothing to talk about.

Which of the following sets contain all of the common factors of 24 and 30?

(A) {1, 2, 5}
(B) {1, 2, 12, 24}
(C) {1, 2, 24, 30}
(D) {1, 2, 3, 6}

Common multiple *A number that is a multiple of 2 or more different numbers; 12 is a common multiple of 2, 3, 4, and 6.*
As Lester watched the 12 amoebas swim underneath a microscope, he admitted that 12 is his favorite number because it's a **common multiple** of 2, 3, 4, and 6 . . . his *other* favorite numbers.

At a circus, 100 kids lined up hoping to win a prize. The clowns plan to give a piece of candy to every 5th child, a balloon to every 10th child, and a poster to every 25th child. How many children will receive all three prizes?

Commutative property *A rule that states changing the order does not change the sum or product.*
Commutative Property of Addition: $a + b = b + a$
Commutative Property of Multiplication: $ab = ba$

It doesn't matter if Chris sees the dinosaur display then the robot display or if he sees the robots then the dinosaurs. The **Commutative Property** states that he'll have fun either way.

Using the commutative property, what number belongs in the answer blank?

$$6 \times 3 = \_\_\_ \times 6$$

Compass *An instrument consisting of 2 legs hinged together that is used to draw circles and arcs.*
Johnny grabbed a **compass** in the inventions gallery and sketched a perfect circle on a sheet a paper.

A compass may be used when constructing which of the following?

(A) Segment bisector (C) Congruent triangles
(B) Angle bisector (D) All of the above

Complementary angles *Two angles whose measures have a sum of 90°.*
Mr. Stickler told the cashier at the café that he wasn't going to actually eat all three chocolate biscotti *himself.* He needed them to show a student how two **complementary angles** equal a sum of 90°.

If the measure of an angle is 40°, find the measure of its complement.

Composite number *A natural number that has more than 2 factors.*

Chris yawned as Lester went on to say that one reason 12 is his favorite number because it is also a **composite number.** It has six factors: 1, 2, 3, 4, 6, and 12.

Which of the following numbers is composite?

(A) 23 (B) 21 (C) 7 (D) 11

Cone *A 3-dimensional figure that has a circular base and 1 vertex.*

Johnny twisted a sheet of paper into a **cone** and plopped it onto his head like a hat, though he had no idea why.

Which figure is a cone?

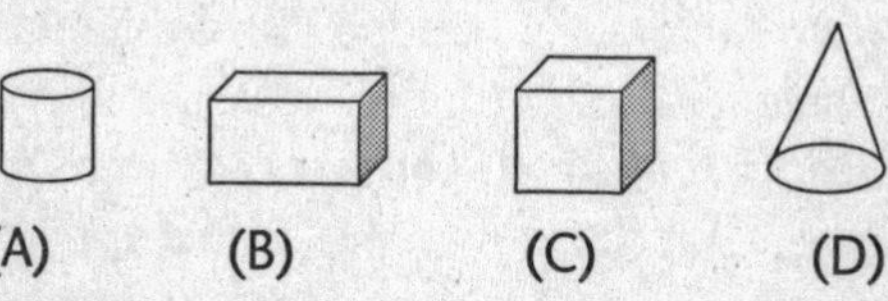

Congruent *Figures that have the same size and shape; indicated by the symbol ≅.*
Sonia was amazed when she found out how snowflakes are formed . . . and that no two flakes are **congruent.**

Which polygon is **congruent** to (≅) △ ?

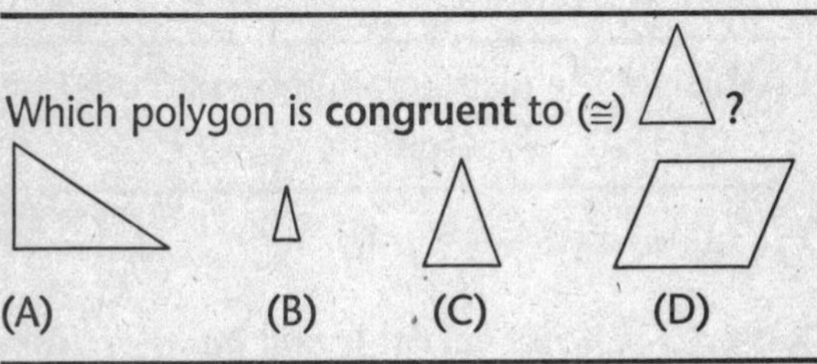

Consecutive integers *Integers in order that follow one after the other; the first 3 consecutive integers are 1, 2, and 3; the first 3 consecutive odd integers are 1, 3, and 5; the first 3 consecutive even integers are 2, 4, and 6.*
When Mr. Stickler handed Wendy, Sonia, and Anna each a biscotti, the girls gave him a quick cheer that included **consecutive** even **integers.** "Two, four, six, eight, who do we appreciate!"

Which of the following sets of numbers are consecutive integers?

(A) 23, 25, 26 (C) 5, 10, 12
(B) 9, 11, 12 (D) 14, 15, 16

Construction ***Using only a compass and straight edge to draw a figure***
Johnny realized that he liked using a compass and a ruler to develop **constructions.** Maybe he'd become an architect some day!

What is the process called when a compass and straight edge are used to draw congruent triangles?

Coordinates ***The position of a point on a graph written as an ordered pair (x, y); the first coordinate is the horizontal position (x-axis), and the second coordinate is the vertical position (y-axis).***
Much to Chris's dismay, Lester insisted on spewing out the **coordinates** of their next exhibit rather than just saying the Human Body Gallery.

Using the following line graph, write the coordinates that represent the number of students enrolled in Williams Middle School in 2006.

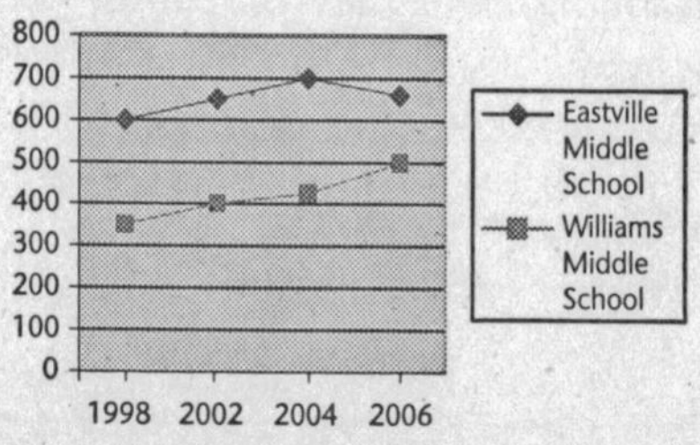

Counting principle *If one event has m possible outcomes and a second event has n possible outcomes, the number of possible outcomes when both events occur is m × n.*
For Example: If Tim rolls a die and tosses a coin, there are 12 possible outcomes because the first event (rolling a die) has 6 outcomes and the second event (tossing a coin) has 2 outcomes.

Chris used the **counting principle** to figure that if Lester turned right at the end of the amoeba display and Chris turned left, there would still be only one outcome. They'd both end up at the Human Body Gallery.

If Gabby has 3 shirts and 2 skirts, how many different outfits can she make using 1 shirt and 1 skirt?

Cross multiplication *In a proportion, the products of the diagonals are equal; if the products are equal, then so are the ratios.*
Hoping to buy a cool book in the museum store, Anna quickly used her **cross multiplication** skills to figure out how much 25% off of $10 was.

Solve $\frac{1}{5} = \frac{4}{x}$

Cube *A solid figure that has 6 congruent square faces, 8 vertices, and 12 edges.*
Anna noticed that the "ice cube" floating in her cup of soda was not a *true* **cube** because its sides were not exactly congruent.

Which of the following figures is a cube?

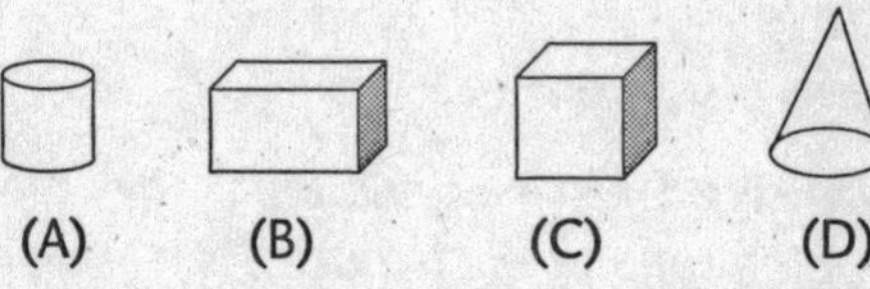

Cubed *A number or quantity raised to the third power; 4^3 is read as 4 cubed and means $4 \times 4 \times 4$.*
Lester admitted that he wished he could **cube** himself so that he could take geometry, algebra, and probability classes all at the same time.

Evaluate 2^3.

Cube root *The cubed root of y is x if x × x × x = y; 4 is the cubed root of 64 because 4 × 4 × 4 = 64;* $\sqrt[3]{64} = 4$.

In certain circles Wendy equaled the "**cubed root** of cool."

Evaluate $\sqrt[3]{8}$.

Cumulative frequency *Obtained by keeping a running total of the frequencies.*

Anna kept a **cumulative frequency** chart that noted the number of times Wendy applied nail polish in a given day. So far, Wendy was up to 15 applications.

Using the table below, complete the cumulative frequency column.

Number of Televisions	Tally	Frequency	Cumulative Frequency
1					3	
2	𝍸	5	
3	𝍸 ||	7	

Customary measurement system *A system that measures:*

- *distance in feet, yards, and miles*
- *weight in ounces, pounds, and tons*
- *volume in cups, pints, quarts, and gallons*
- *temperature in degrees Fahrenheit*

Mr. Stickler could not explain why the United States uses the **customary measurement system** while the rest of the world uses the metric system.

Convert 3 yards to feet.

Cylinder ***A solid shaped like most cans; it has two equal, parallel bases that are congruent circles.***
Why does this always happen to me? Lester wondered as he stared at the empty toilet paper **cylinder** in the bathroom stall.

Which of the following figures is a cylinder?

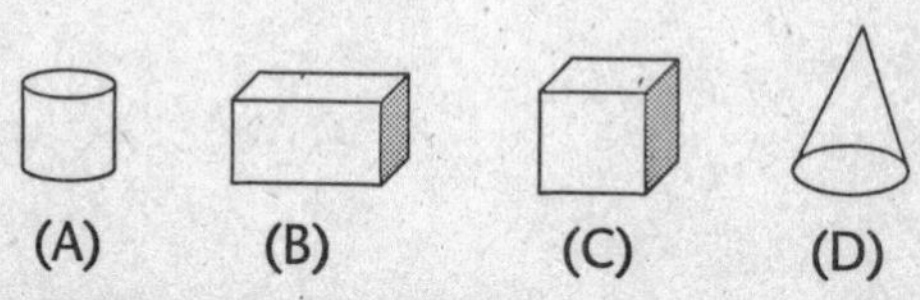

Data *Information gathered for statistical purposes.*

Mr. Stickler asked his students to try some of the hands-on experiments in the Weather Gallery and to record the **data.**

Mr. Stickler's 15 students were asked to write down the number of pets that live in their home. Their answers are written below. Record this set of data in a tally chart.

2, 3, 3, 1, 3, 1, 1, 2, 3, 2, 2, 3, 3, 3, 2

Number of Pets	Tally	Frequency

Decimal ***The numbers in the base 10 number system, having 1 or more places to the right of a decimal point. A number with a decimal point can be converted to a fraction or percent.***
Gabby admitted that she would not buy the dinosaur-shaped pen because it had a **decimal** on the price tag and she did not want to be stuck carrying loose change.

Compare these decimals: 1.43, 1.4, and 1.5. Put them in order them from least to greatest.

Degree ***The unit of measure of an angle or arc; 1°, read as 1 degree, is equal to 1⁄360 of one complete rotation; a circle contains 360°; a triangle contains 180°.***
Lester realized that he was headed in the wrong direction and made 180 **degree** turn toward the tornado machine.

Choose the best estimate of the angle's measure.

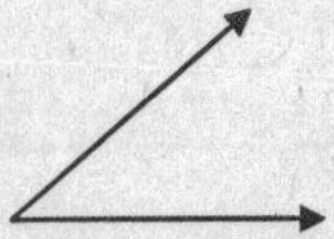

(A) 5° (B) 40° (C) 90° (D) 180°

Denominator *The bottom number or expression in a fraction; in the fraction, $\frac{5}{7}$, 7 is the denominator and 5 is the numerator.*
Anna felt like the **denominator** in a fraction when Lester accidentally fell and squashed her to the ground.

Using the chart below, Sonia said $\frac{7}{20}$ of the children in her class have brown eyes. What is the denominator of this fraction? What does the denominator represent?

Eye Color	Number of Students
Blue	10
Brown	7
Green	3

Diagonal ***A line segment connecting 2 nonconsecutive vertices of a polygon.***
Lester did not mean to laugh at Anna but he realized that a **diagonal** of snot had emerged from her left nostril and stretched all the way to her eyebrow.

In the following diagram, which of the following segments is a diagonal?

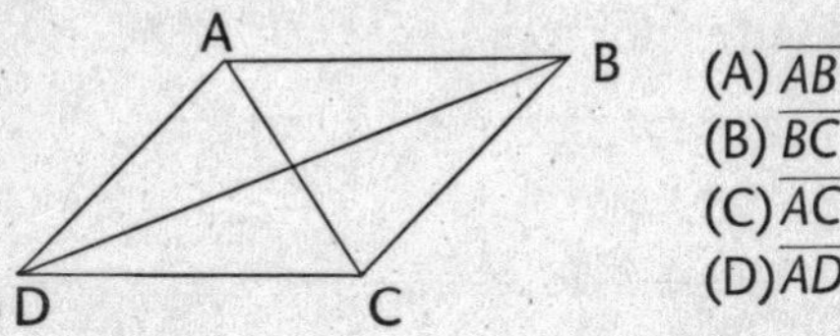

(A) $\overline{AB}$
(B) $\overline{BC}$
(C) $\overline{AC}$
(D) $\overline{AD}$

Diameter ***A line segment with endpoints on the circle that also passes through the center of the circle; the diameter is twice the length of the radius.***
Angry, Anna told Lester that the **diameter** of his brain was equal to that of a pea.

Which segment in the following diagram is a diameter of circle F?

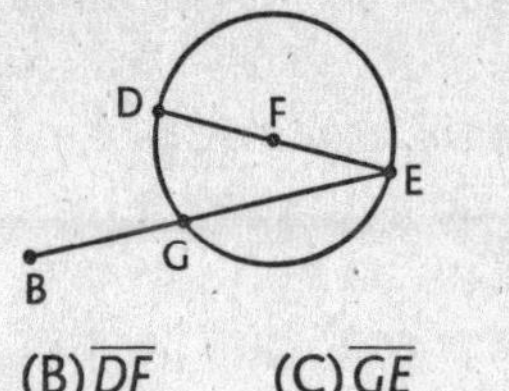

(A) $\overline{DE}$ (B) $\overline{DF}$ (C) $\overline{GE}$ (D) $\overline{BE}$

Difference ***The result of subtraction.***
When Wendy subtracted the cost of her museum snack from the $20 in her pocket, she was left with the **difference** of $15.

Find the difference between 52 and 18.

Digits ***The symbols 1, 2, 3, 4, 5, 6, 7, 8, and 9 that are used to write numbers.***
Chris wondered if it would be cooler to ask Gabby for her "**digits**" rather than her number.

Which digit in the number 36.4921 is in the tenths place?

Discount ***A percentage that is subtracted from the original price of an item.***

Example: If a \$20 shirt is discounted 10%, the discount is 10% of 20 = 10% × 20 = 0.10 × 20 = \$2; consequently, the shirt costs \$20 – \$2 = \$18.

Gabby bought a tie for her dad that had tiny Albert Einstein heads all over it not because it was cool, but, because it was **discounted** 70%.

If all shoes are now 25% off at Molly's Boutique, what is the sale price for a pair of shoes that originally costs \$40?

Distributive property ***States that for any real number a, b, and c: a(b + c) = ab + ac.***

3(4 + 5) = 3(4) + 3(5) = 12 + 15 = 27; 4(a – 5) = 4a – 4(5) = 4a – 20

The **distributive property** states that Lester would end up with the same amount of popcorn whether he bought two servings in a big bowl or if he bought it in two individual-sized bags.

Compute 5 × 98 using the distributive property [hint: 5 × 98 = 5(100 – 2)].

Divide (÷) *To split into equal groups.*
Low on cash, Lester decided to buy one small bag of popcorn and **divide** the kernels equally between Anna and him.

Compute 360 ÷ 6.

Dividend *The value being divided in a division problem; dividend ÷ divisor = quotient.*
Counting up all the popcorn kernels before splitting them, Lester came to the conclusion that he had a **dividend** of 244.

Which number in the equation 24 ÷ 4 = 6 is the dividend?

Divisible *Leaves no remainder when divided by a number; 12 is divisible by 3 because 12 ÷ 3 = 4.*
Good thing 244 popcorn kernels was **divisible** by two. Otherwise who would get to eat the remainder?

Is 42 divisible by 5? Why or why not?

Divisor *The value that divides another value in a division problem; dividend ÷ divisor = quotient.*

Lester sure was glad that he had only a **divisor** of 2 when it came to his 244 kernels of popcorn because 122 kernels was barely enough for three handfuls.

In the equation 24 ÷ 4 = 6, which number is the divisor?

E

Element *A member of a set.*

Anna said "thanks but no thanks" when Lester pushed her **element** of popcorn across the table. She didn't want to eat it after he'd spent an hour touching it and licking his fingers.

How many elements are in the set below?

{3, 6, 9, 12, 15}

Endpoint ***A line segment contains 2 of these points that each mark the end of the segment.***

Gabby groaned, exhausted, when she realized that Sonia wasn't even close to the **endpoint** of her story about magnets.

A line segment is named using its endpoints. What is the name of the line segment below?

Equals ***A mathematical relationship used to show two quantities that have the same value.***
Anna and Wendy were total **equals** when it came to the last math test. They both scored a 98.

If $x + 3 = 7$, what is the value of x?

Equation ***A mathematical sentence stating that 2 quantities are equal.***
Gabby and Sonia don't know it but they are so alike, they're like two sides to an **equation.**

Which of the following is an equation?

(A) $x + 3$

(B) $x - 5 < 12$

(C) $3 + 5 = 10$

(D) $x - 4 = 8$

Equilateral triangle ***A triangle with 3 equal sides and 3 equal angles (all angles measure 60°).***
Johnny wondered how the ancient Egyptians made their pyramids into **equilateral triangles** without the aid of machinery.

Are all equilateral triangles similar?

Equivalent fractions *Two fractions having the same value;* $\frac{3}{4}$ *and* $\frac{6}{8}$ *are equivalent.*
Gabby and Sonia are much like **equivalent fractions.** On the surface they look like complete opposites but, with a little factoring, they soon come to realize that they are exactly the same.

List three fractions that are equivalent to $\frac{1}{2}$.

Estimate *Making an educated guess based on the given information; an approximation or rough calculation.*
Chris glanced at his watch when Lester disappeared into the boys' room and **estimated** that, if he left now, he could ditch his "museum buddy" for the rest of the trip.

Estimate: 3.72 + 2.19 + 9.06 + 4.97.

Even number *A number that is divisible by 2; even numbers end in a 0, 2, 4, 6, or 8.* Lester remarked to his museum buddy that it took only 128 running steps to catch up. He also added that 128 happened to be an **even number.**

Which of the following is not an even number?

(A) 10 (C) 47

(B) 28 (D) 52

Exponent *Shows the number of times a number (or quantity) should be multiplied by itself; in the expression 5^3, 3 is the exponent; $5^3 = 5 \times 5 \times 5 = 125$.*
Johnny thought he had 5 bucks in his pocket, so he was psyched to find that he really had 5 bucks with an **exponent** of two in his jeans. . . a.k.a., 25 bucks!

Which of the following is equal to 100,000?

(A) 10^4 (C) 10^6

(B) 10^5 (D) 10^7

Expression *A mathematical phrase that contains numbers, variables, and operations; an expression does not contain an equal sign or an inequality symbol.*

Anna wandered into the Calculus Center by accident, and she quickly exited because she didn't understand any of the mathematical **expressions.**

Which of the following is a mathematical expression?

(A) $2 + 7$ (C) $9 + 2 \neq 10$

(B) $5 < 7$ (D) $12 - 4 = 8$

Exterior angle of a polygon *An angle on the outside of a polygon created by extending a side of a polygon.*

Johnny asked Chris why he was hiding within the **exterior angle** of the door to the Crystals display, and Chris explained that he was hiding from Lester.

In the diagram below, which of the following angles is an exterior angle of ΔBCD?

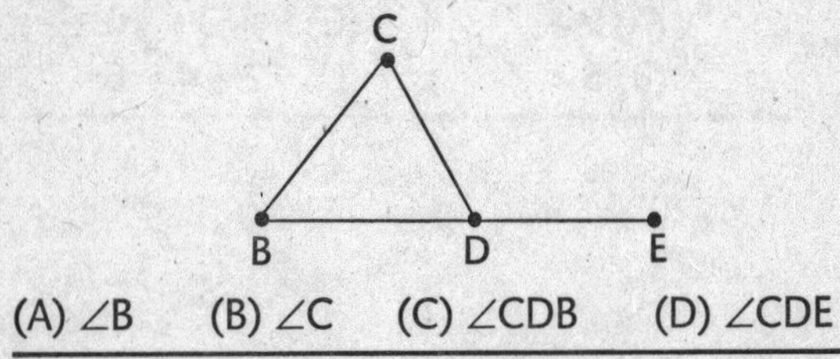

(A) $\angle B$ (B) $\angle C$ (C) $\angle CDB$ (D) $\angle CDE$

Face *One of the flat surfaces of a three-dimensional figure.*
Anna touched the smooth **face** of the purple crystal on display in the rocks and resources gallery.

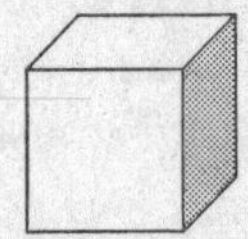

How many faces does the figure above have?

Factor ***A whole number that when divided into another number leaves no remainder.***
Buying yet another snack, Mr. Stickler told the cashier at the museum café that he'd only eaten 2 of the many biscotti he'd bought that day, which was a **factor** of the 22 he had really eaten.

List all of the factors of 20.

Factorial *The product of a whole number and each of the natural numbers less than that number; $5! = 5 \times 4 \times 3 \times 2 \times 1$.*

Munching on his chocolate snack, Mr. Stickler realized that the biscotti was so delicious that he'd probably eat the **factorial** of 22 cookies if the café had that many to sell him.

Evaluate 4!

Factoring *Breaking the number (or quantity) into the product of other numbers (or quantities).*

Factoring in all the possibilities of where his museum buddy should be, Lester came to the conclusion that Chris must be lost.

Use your factoring skills to list the factors of 32.

Fahrenheit *A way to measure temperature in the customary unit system; 32° Fahrenheit (F) is the freezing point of water; 212° F is the boiling point of water.*
Johnny pulled off his extra sweater when he entered the greenhouse-effect room because it felt like it was 90° Fahrenheit in there.

Using the formula $F = \frac{9}{5}C + 32$, where F is Fahrenheit and C is Celsius, convert 10° Celsius to Fahrenheit.

FOIL *An acronym used to recall the steps for multiplying two binomials; F (multiply the first terms in each binomial) O (multiply the outer terms), I (multiply the inner terms), L (multiply the last terms) in each binomial.*
For Example: $(x + 3)(x + 5) = x^2 + 5x + 3x + 15$
$= x^2 + 8x + 15$
Sonia was relieved to find that the All Things **Foil** collection contained cool stuff made from aluminum and had nothing to do with multiplying two binomials.

Multiply 101 × 101 using the FOIL method.
[Hint: 101 × 101 = (100 + 1)(100 + 1).]

Formula *A mathematical rule that helps solve problems.*
Mr. Stickler frantically searched for a pen and paper when he'd spotted the **formula** for making cotton candy in the Chemistry Corner.

Write the formula for the area of a triangle.

Fraction *Part of a whole when the whole is divided into equal sections; if a candy bar is cut into 4 equal pieces and you eat one piece, you have $\frac{3}{4}$ of the candy bar left.*
Sonia felt like a total **fraction** when Wendy and Gabby chose to continue making their own robots rather than come to the bathroom with her.

What fraction of the shapes shown below have five sides?

Frequency ***The number of times an item appears in a set.***

Anna wondered if Mr. Stickler had a weakness for sweets as she noticed the man wolfing down cookies with alarming **frequency.**

The data below shows the number of pets Mr. Stickler's students have in their homes. If a student from Mr. Stickler's class is chosen randomly, what is the probability the student chosen will own 3 pets?

Number of Pets	Tally	Frequency
1					3
2	~~||||~~	5
3	~~||||~~ ||	7

Gallon *A customary unit of capacity; 1 gallon = 4 quarts = 8 pints = 16 cups.*

Johnny was amazed that he had constructed a working toy sailboat with only an empty **gallon** jug of milk, a few Popsicles, and a piece of paper.

Fill in the blank: 5 gallons = _____ pints.

Gram *A base unit in the metric system that measures mass; 1 kilogram = 1,000 grams.*

Chris placed his plastic pen cap onto a metric scale and learned that it weighted exactly 1 **gram.**

Which is larger: 1 kilogram or 1 gram?

Greater than ***Symbol (>) that shows the first term is larger than the second.***

Lester marveled at the gigantic ball of aluminum in the All Things Foil collection because it was at least ten times **greater than** the one he had in his room at home.

Which of the following makes a true statement?

(A) $7 > 10$ (C) $-6 > -2$

(B) $5 > 3$ (D) $0 > 2$

Greater than or equal to ***Symbol (≥) that shows the first term is larger than or equal to the second.***

Sonia said she'll split the check at the museum restaurant equally only if her meal in is **greater than or equal** to the cost of friends' food. Otherwise she wants to pay separately.

Which of the following makes a true statement?

(A) $-6 \geq 5$ (C) $0 \geq 8$

(B) $4 + 3 \geq 4 + 5$ (D) $2 \times 3 \geq 1 \times 5$

Greatest common factor *The largest factor of 2 or more numbers (or quantities).*

The **greatest common factor** between Wendy, Gabby, and Sonia is that they all live on the same street.

Find the greatest common factor of 28 and 42.

Height (altitude) ***The perpendicular distance between a vertex and the opposite side opposite of the polygon.***

The global warming gallery showed Anna that the **height** of the glaciers in Antarctica was getting shorter and shorter every year.

In the diagram shown below, which line segment is an altitude?

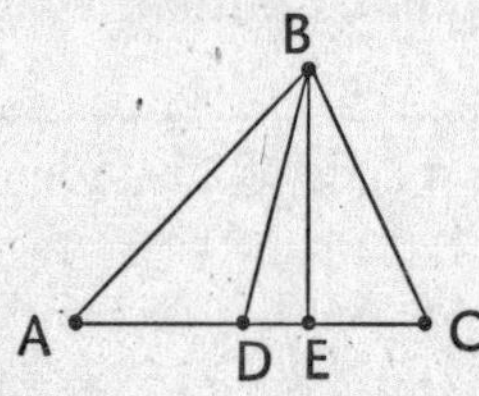

(A) $\overline{BA}$ (C) $\overline{BE}$

(B) $\overline{BD}$ (D) $\overline{BC}$

Hexagon *A 6-sided polygon.*

Lester liked the **hexagon**-shaped table in the experiment gallery because 6 people could work there at the same time.

Which figure is a hexagon?

(A) (B) (C) (D)

Histogram ***A graph that shows the data from a frequency table with blocks of proportionately sized areas.***

Mr. Stickler was surprised that Chris arrived at the cafeteria on time because, according to Mr. Stickler's **histogram,** the student was usually late.

Test scores for a class of 20 students are given as:

72, 100, 95, 65, 86, 88, 76, 89, 68, 79, 84, 72, 77, 67, 80, 71, 83, 73, 78, 90

This data is recorded into a table.

Test Scores	Frequency
61–70	3
71–80	9
81–90	6
91–100	2

Then the data is displayed in the histogram.

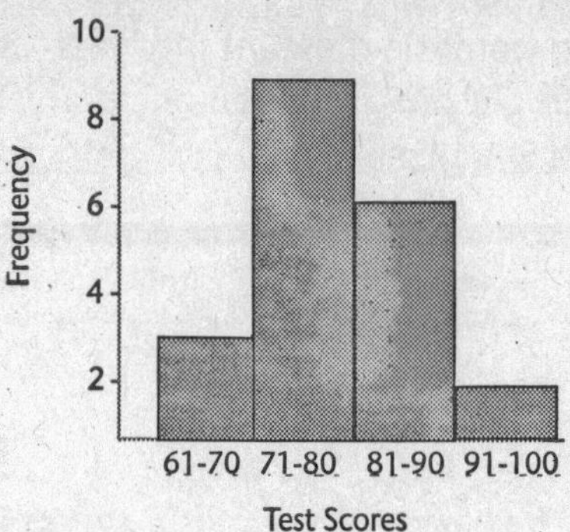

The histogram below shows the results of 20 students who took their biology final exam.

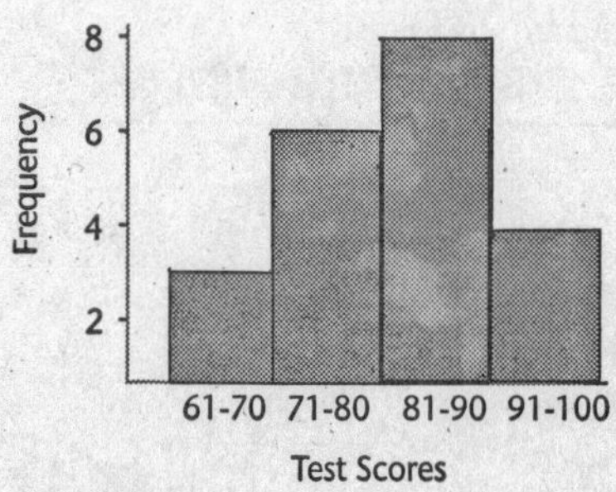

How many students received a score between 91 and 100?

(A) 3 (B) 4 (C) 6 (D) 8

Horizontal ***A line parallel to the horizon; the floor is horizontal.***

A bit tired from the museum madness, Chris decided to get **horizontal** on one of the cafeteria's picnic tables.

Is the *x*-axis vertical or horizontal?

Hundreds ***The place value that is 3 spaces to the left of the decimal point.***

Anna suggested that Lester turn in the **hundreds** of dollars he'd discovered in the aviation gallery to the Lost and Found.

Which digit is in the hundreds place in the number 4,367.28?

(A) 3 (B) 6 (C) 2 (D) 8

Hundredths ***The place value 2 spaces to the right of the decimal point.***

Lester explained that he'd found only $\frac{1}{100}$ of a dollar, and nobody was going to be upset about losing a penny.

Which digit is in the hundredths place in the number 4,367.28?

(A) 3 (B) 6 (C) 2 (D) 8

Hypotenuse ***The side opposite the right angle in a right triangle; the hypotenuse is the longest side of a right triangle.***

Gabby cut her grilled-cheese sandwich into two right triangles. Then she picked up one and bit into its **hypotenuse.**

Name the hypotenuse in the triangle below.

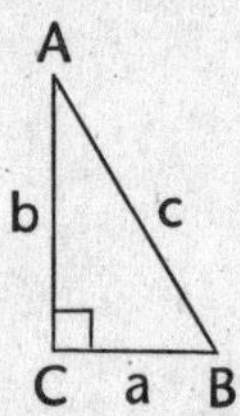

I

Identity property *The number that, when combined with another number under a given operation, does not change that number; zero is the additive identity because $x + 0 = 0 + x = x$; One is the multiplicative identity because $x \times 1 = 1 \times x = 1$.*

To some, Wendy's reading glasses might be considered an **identity property** because, although they make her *look* smarter, wearing them does not improve her test scores.

Give 3 examples showing zero as the additive identity.

Improper fraction *A fraction with the numerator larger than the denominator; the value of an improper fraction is always greater than one; 4⁄3 is an improper fraction.*

Sonia thought Lester kind of looked like an **improper fraction** because, unfortunately, his head is way too big for the rest of his body.

Which of the following fractions is improper?

(A) $\frac{1}{2}$ (B) $\frac{7}{4}$ (C) $\frac{2}{5}$ (D) $\frac{5}{9}$

Increase *To get bigger.*

Gabby noticed a strange **increase** in flies around her when she sprayed on her new strawberry-scented perfume.

Using a mathematical operation, write the expression 4 increased by 3.

Inequality *Symbols (<, >, ≤, ≥, or ≠) used to state a relationship between expressions.* Anna wanted to know why there was always a line at the girls' bathroom and not at the boys'. Was there an inequality in the number of girls' versus boys' rooms at the museum?

Simplify $2 + 4 \times 3 < \frac{20}{2} + 3 \times 5$.

Infinity *Symbol (∞); too large to be measured; infinity is larger than any number you can imagine.* Wendy wished that she had never entered the human body gallery when she heard about the **infinite** number of creepy crawly organisms living on her skin.

Is 9,215,978,999,999 bigger than ∞?

Integer *the set of positive and negative whole numbers and zero {. . . –2, –1, 0, 1, 2, . . .}.* Although Lester was surprised that he'd rated an integer of 3 on Sonia's cool personality index, he was not upset.

Which of the following is not an integer?

(A) $\frac{20}{5}$ (B) 4.2 (C) –7 (D) 8.0

Inverse *An additive inverse is a number that when added to a given number equals 0; (the additive identity); a multiplicative inverse is a number that when multiplied by a given number equals 1 (the multiplicative identity).* When asked why scoring 3 on a coolness test didn't bother him, Lester remarked that it could have been worse. He could have gotten the additive **inverse** of three: –3.

Find the additive inverse of 4.

Irrational *A real number that cannot be expressed as the ratio of 2 integers (denominator not 0); irrational numbers are nonrepeating, nonterminating decimals.*

Lester's appetite for knowledge was like an **irrational** number with nonterminating decimals. It never ended.

Which of the following is an irrational number?

(A) 4.2 (C) $\sqrt{100}$

(B) –5.0 (D) $\sqrt{7}$

Isolate (isolation) *The process of solving for a variable; to get the variable by itself.*

As any good friend would do, Wendy **isolated** Gabby after lunch and gave the girl a stick of gum for her onion breath.

Solve for x in the equation $x - 2 = 10$.

Isosceles trapezoid *A trapezoid whose legs and base angles are equal.*

Wendy thought it a bit strange when Lester cut his peanut butter-and-jelly sandwich into an **isosceles trapezoid,** whose legs and base angles were perfectly equal.

Which of the following is an isosceles trapezoid?

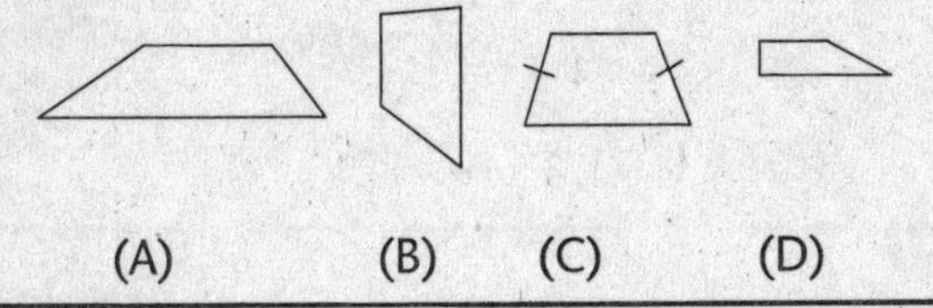

(A) (B) (C) (D)

Isosceles triangle *A triangle with 2 equal sides and 2 equal angles.*

Sonia suggested that they change their special secret greeting from forming an acute triangle with their pinkies to making an **isosceles triangle** with their index fingers.

Which of the following triangles is isosceles?

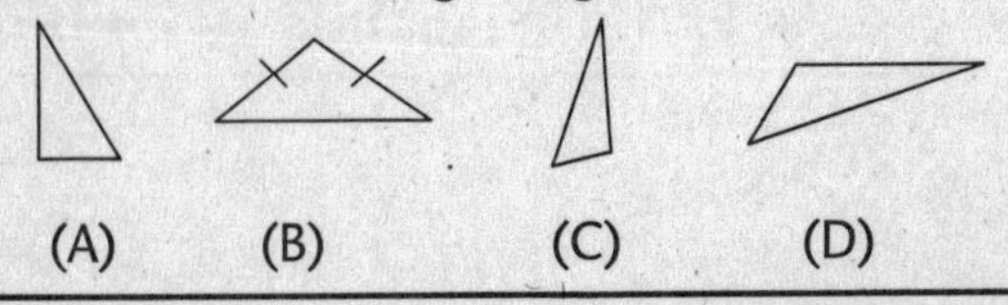

(A) (B) (C) (D)

L

Least common denominator ***When adding or subtracting fractions, the least common denominator (LCD) is the smallest number that can be used for both denominators; it is the least common multiple of the original denominators.*** Not unlike finding the least common denominator in fractions, Wendy tried and tried until she discovered something that Sonia and Gabby had in common: They both loved the 3-D Aquatic Adventure at the science museum.

Simplify $\frac{1}{2} + \frac{1}{6}$.

Least common multiple ***The smallest multiple of 2 or more integers.***

Wendy believed that finding a good friend was much like finding a **least common multiple.** You have to use trial and error until you get one that fits.

Find the least common multiple (LCM) of 5, 6, and 10.

Leg ***In a right triangle, either side opposite an acute angle; in a trapezoid, either of the non-parallel sides.***

Embarrassed, Johnny said that he was just scratching the bottom **leg** of the triangle on his face when he was really picking his nose.

Name the two legs in the right triangle below.

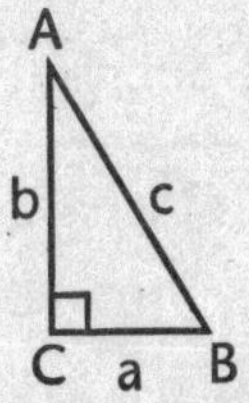

A

b c

C a B

Length *The distance along a line segment; one dimension in a two-dimensional or three-dimensional shape.*
Lester ran across the **length** of the activity center to join his museum buddy, Chris, in an air-and-vacuum experiment.

In the rectangle below, how much longer is the length than the width?

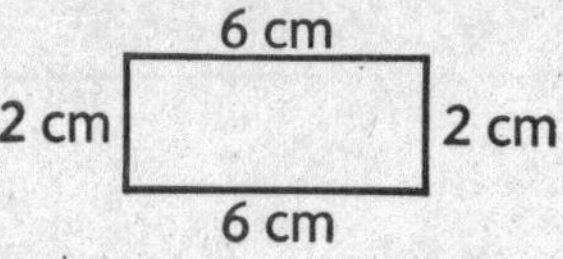

Less than *Symbol (<) that shows the first term has a smaller value than the second term.*
Chris was **less than** happy when he realized that Lester had spotted him.

Compare the numbers, using the symbol > or <: 38 and 271.

Less than or equal to *Symbol (≤) that shows the first term has an equal or smaller value than the second term.*

Feeling a bit insecure, Anna would join kids at an activity table if their intelligence seemed to be **less than or equal to** hers.

Place this group of numbers in order from **least to greatest** using inequality symbols: 567, 576, 509, 576, and 5,670.

Like terms *Terms that contain the same variables and corresponding exponents; terms must be "like terms" in order to be combined by addition or subtraction.*

Anna couldn't believe that she and Chris had so many **like terms.** They are both excellent swimmers, both hang out at the same after school program, and both think the dinosaurs are the coolest exhibit at the museum.

Which of the following pairs are like terms?

(A) $2x$ and $2y$

(B) $4x$ and 5

(C) x^2 and x^3

(D) $5x^2$ and $7x^2$

Line *A path extending in both directions connecting an infinite number of points; line AB can be written as* $\overleftrightarrow{AB}$, $\overleftrightarrow{BA}$, *or by a single lowercase letter that labels the line.*

Johnny felt sorry for the girls waiting to use the bathroom because the **line** seemed to go on forever.

Give at least 3 different names for the line below.

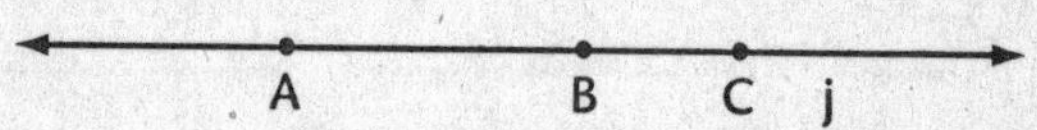

Line graph *A graph that uses connected line segments to display data showing change over time.* Mr. Stickler added 22 biscotti to his **line graph** that charted his weekly biscotti eating.

Using the line graph below, how many people would you expect to be swimming at the town pool on June 17?

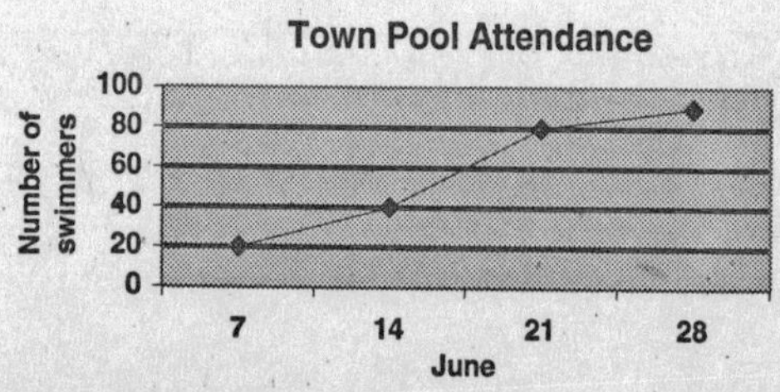

Line segment *Part of a line; the endpoints of line segment $\overline{AB}$ (also called $\overline{BA}$) are points A and B.*

Mr. Stickler's asked his class to formed a **line segment** outside of the famous eco-house as they waited their turn to enter it.

Give at least 3 different names for the line segment below.

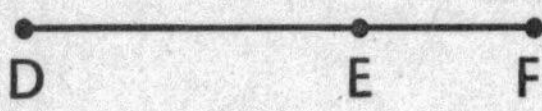

Line of symmetry *A line on which a figure can be folded in half so both halves are identical.*

Lester noticed that Anna formed a **line of symmetry** in between her eyebrows every time she concentrated on something.

A heart has one line of symmetry.

Which of the following letters does not have line symmetry?

(A) T (B) B (C) M (D) F

Linear equation ***An equation that describes a straight line; all nonvertical lines are written in the form $y = mx + b$.***

Mr. Stickler plugged his latest 22 biscotti into a **linear equation** that, when graphed, made a line that showed the more he ate, the heavier he got.

Name the point where lines j and k intersect.

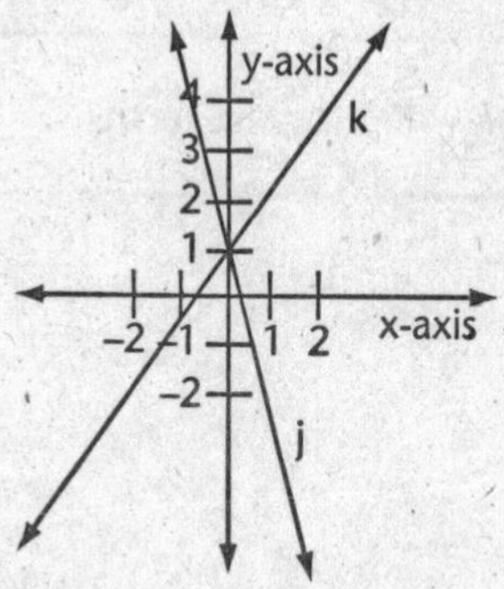

Liter
A base unit in the metric system that measures capacity; 1 liter = 1,000 milliliters.

Johnny drank a whole **liter** of water after running for a half hour on the stress-test treadmill in the Human Body Gallery.

Fill in the blank: 0.01 Liters = ___Milliliters.

Lowest terms *When the greatest common factor of a numerator and denominator is 1.*

The cashier told Mr. Stickler that a pack of gum cost $\frac{25}{100}$ dollars, but when he reduced the fraction to its lowest terms he figured out that she really meant 25¢.

Write $\frac{5}{15}$ in lowest terms.

M

Mean (average) *A value that is found when the sum of a set of numbers is divided by the number of values in the set.*

The **mean** age of the students in Mr. Stickler's class is 11.

The hours spent by five students in completing a math project are 10, 8, 5, 9, and 18. What is the mean of the number of hours?

Median ***The value that falls in the middle of the set when the numbers are written in order; if there is an even number of values, take the average of the 2 middle numbers.***

Wendy narrowed down her souvenir shopping to 3 mugs and then picked the mug with the **median** price.

Which statement is true for the set of numbers 6, 7, 11, 6, 8?

(A) Mean > mode.

(B) Median > mean.

(C) Mode > median.

(D) Mean = median.

(E) Median = mode.

Meter ***A base unit in the metric system that measures length.***

1 kilometers = 1,000 meters

1 meter = 100 centimeters = 1,000 millimeters

Wendy asked Gabby to stay at least 2 **meters** away from her because, despite the gum, Gabby's breath reeked.

Fill in the blank: 50,000 cm = ______ km

Metric system *A base 10 system of measurement in which:*

- *Length is measured in millimeters, centimeters, meters, and kilometers.*
- *Mass is measured in grams and kilograms.*
- *Volume is measured in milliliters and liters.*
- *Temperature is measured in degrees Celsius.*

+1000	+100	+10	Base Unit	× 10	× 100	× 1000
$\frac{1}{1000}$Kilo-	$\frac{1}{100}$Hecto-	$\frac{1}{10}$Deka-	1 Meter	10 Deci-	100 Centi-	1000 Milli-
			1 Gram			
			1 Liter			

Wendy told her friends that she's moving to Paris just as soon as she learns how to convert American measurements to the **metric system** . . . and learns how to speak French.

Place in order from least to greatest:
1 meter, 1 millimeter, 1 centimeter, and 1 kilometer.

Midpoint ***The point on a line segment that divides it into 2 equal parts.***

Lester wished that he could stand in the **midpoint** of the line because that would place him right behind Anna.

If point H is the midpoint of $\overline{GI}$, which segments must be equal?

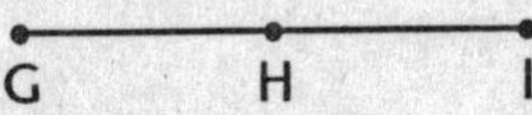

Mile ***A customary unit of length; 1 mile = 5,280 feet.***

After 30 minutes on the treadmill, Johnny was surprised to learn that he had run only 5,280 feet, a.k.a. one **mile.**

Fill in the blank (Hint: 3 feet = 1 yard):
1 mile = ______ yards.

Minus (–) ***The symbol for subtraction.***

Sophia's favorite exhibit at the science museum was the Human Body Gallery . . . **minus** the blood and guts.

Compute 63 minus 41.

Mixed number *The sum of a whole number and a proper fraction;* $3\frac{1}{2}$, $-1\frac{3}{4}$, $5\frac{1}{7}$.

Mr. Stickler munched on another half of biscotti, turning his total number eaten to 22½, which happens to be a **mixed number.**

Write $\frac{9}{4}$ as a mixed number.

Mode *The number that occurs most often in a set; if there are 2 values that appear most often, the set is bimodal, and if no value appears, most often the set has no mode.*

Gabby's **mode** of lunch is egg salad with celery and onions because she eats it every day except Tuesday, when she has tuna fish with celery and onions.

Find the mode of the given set: 88, 90, 91, 78, 91, 88, 91.

Monomial *A polynomial with one term; $4x^2y^3$, $-5x$, and y are three monomials.*
Having been ditched by his "museum buddy," Chris, Lester felt like a **monomial** in a museum full of polynomials.

Which of the following is a monomial?

(A) $3x^2$ (C) $3x^2 + 4x - 1$
(B) $3x + 4$ (D) $4x + 3y$

Multiple *The product of a number and a whole number; multiples of 6 are 6, 12, 18, 24, 30, 36, and any number that can be divided evenly by 6.*
Needing smaller bills, Chris asked Mr. Stickler if he had five \$20 bills for one \$100 bill because 100 is a **multiple** of 20.

List the first 5 multiples of 3.

Multiply (× or ·) *Adding a number to itself a given number of times; $3 \times 6 = 18$ or $6 + 6 + 6 = 18$.*
Mr. Stickler quickly **multiplied** \$20 times 5 in his head to make sure it equaled \$100.

Compute 15×3.

N

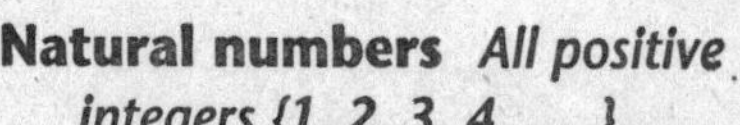

Natural numbers *All positive integers {1, 2, 3, 4, . . .}.*

The Creatures We Can't See exhibit explained how to avoid getting sick by counting to 10, using **natural numbers,** while washing your hands with soap and water.

Which of the following is a natural number?

(A) –4 (B) 0 (C) 7.6 (D) 1,245

Negative numbers *A number that is less than 0.*

Technically, Johnny had a **negative** amount of money in his pocket because he owed Lester 3 bucks.

Place –3, 4, –5, and 1 on the number line below.

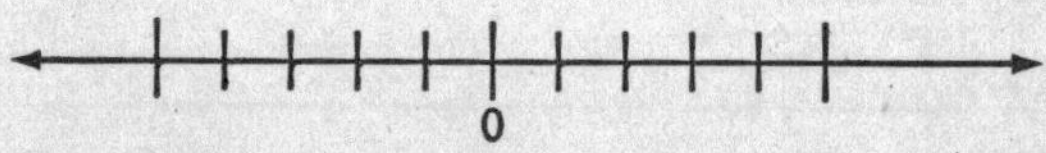

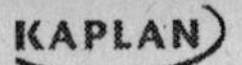

Number line *A line showing the location of all real numbers.*

Johnny plotted the amount of money he had by marking on a **number line.** Unfortunately, it was –3 dollars.

Use the number line to help you round 6,438,689 to the nearest million.

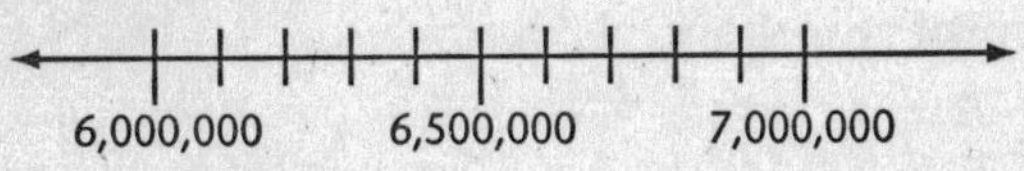

Numerator *The number or expression in the top part of a fraction; in the fraction $\frac{3}{4}$ 3 is the numerator and 4 is the denominator.*

Lester kind of felt like a **numerator** when he accidentally tripped and squashed Chris to the ground.

Using the table below, Sonia said $\frac{7}{20}$ of the children in her class have brown eyes. What is the numerator of this fraction? What does the numerator represent?

Eye Color	Number of Students
Blue	10
Brown	7
Green	3

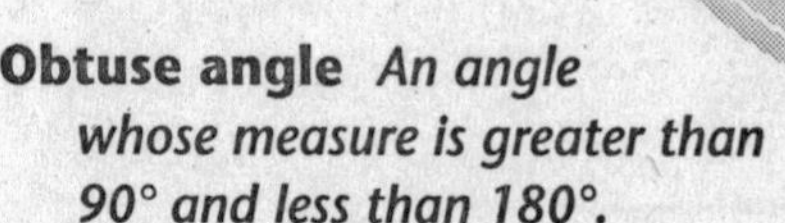

Obtuse angle *An angle whose measure is greater than 90° and less than 180°.*

Chris thought it was awesome that that a T-rex could crack open its mouth to an **obtuse angle.**

Which of the following angles is an obtuse angle?

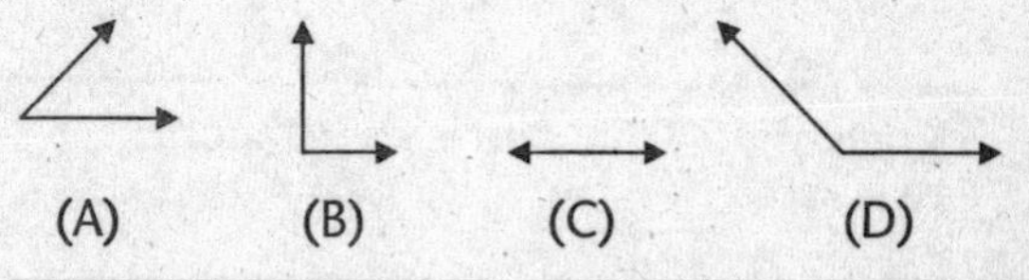

Obtuse triangle ***A triangle with one obtuse angle.***

Wendy folded a square of paper diagonally in the Geometry Genius gallery, forming an **obtuse triangle.**

Which of the following represents the measures of the angles of an obtuse triangle?

(A) 30°, 60°, and 90° (C) 50°, 60°, and 70°
(B) 25°, 45°, and 110° (D) 22°, 78°, and 80°

Octagon ***An 8-sided polygon.***

Once in the Geometry Genus gallery, Lester stood at a red **octagon**-shaped table that reminded him of a stop sign.

Which of the following is an octagon?

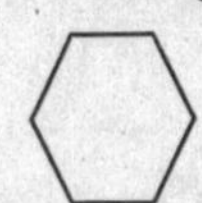

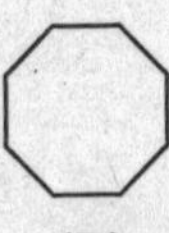

(A) (B) (C) (D)

Odd number ***A number that is not divisible by 2; odd numbers end in a 1, 3 ,5, 7, or 9.***
Lester told the girl next to him that he is 11 years old, which happens to be an **odd number** —not because it's weird, but because it's not divisible by 2.

Which of the following is an odd number?
(A) 41 (B) 52 (C) 88 (D) 130

Ones place ***The place value that is 1 space to the left of the decimal point.***
Johnny reached into his jacket pocket and pulled out a bunch of **ones** that he'd totally forgotten about! Too bad they weren't tens.

What digit in the number 5,692.48 is in the ones place?

Operations *The 4 basic operations used in arithmetic are addition, subtraction, multiplication, and division.*

The scientist in the Healthy Heart section was horrified when Gabby admitted to performing **operations** everyday . . . until she added that *her* operations were addition, subtraction, multiplication, and division.

Which operation is used when finding the sum of 6 and 10?

Order of operations *The correct order in which operations should be performed in an expression; a good way to remember the correct order is PEMDAS, which stands for parentheses first, then exponents, multiplication and division next (left to right), and finally addition and subtraction (left to right).*

Anna has a certain **order of operations** when it came to visiting an exhibit: First she quickly previews the whole exhibit. Then she looks at the individual pieces in detail. Then she picks her favorite one.

Simplify $12 + 5 \times 2$.

Origin *The point (0, 0) where the x-axis and the y-axis intersect on the coordinate plane.*

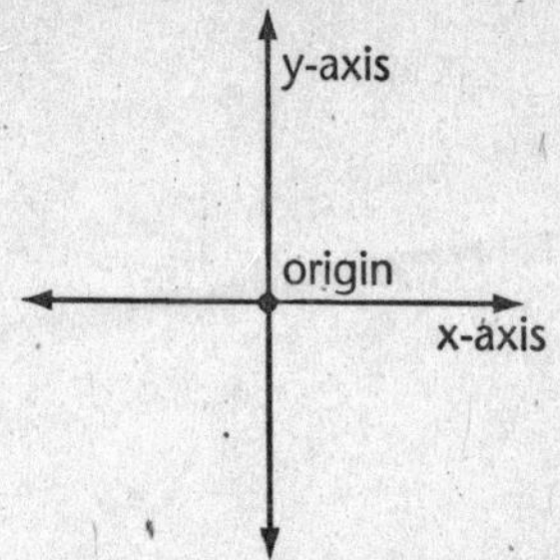

Whoa. Johnny learned at the caveman exhibit that man's **origin** was somewhere in Africa . . . *not* in New Jersey.

Outcome *One of the possible results of an experiment.*

Lester was determined to at least try building his own robot, no matter what the **outcome.**

Find the number of outcomes when a coin is tossed.

Parallel *Lines in the same plane that will never intersect.*

As a joke, Chris stuck a straw in each of his nostrils and ran around the Walrus display with two parallel "tusks" sticking out of his nose.

Which 2 lines in the diagram are parallel?

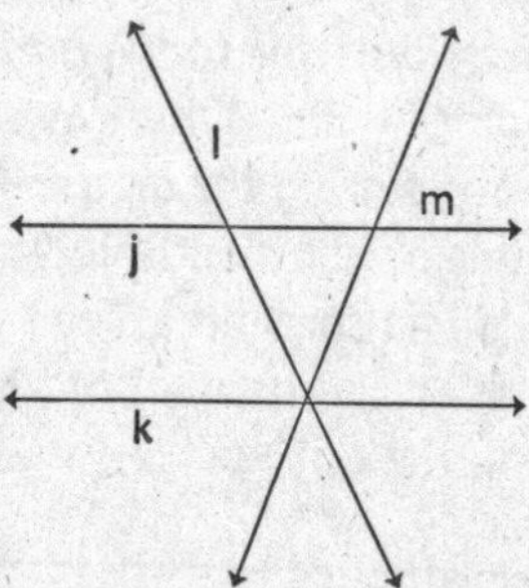

Parallelogram *A quadrilateral with opposite sides parallel and congruent and opposite angles congruent.*
Wendy wondered if anyone ever noticed how her rectangle-shaped purse made the perfect **parallelogram.**

Which of the following figures is a parallelogram?

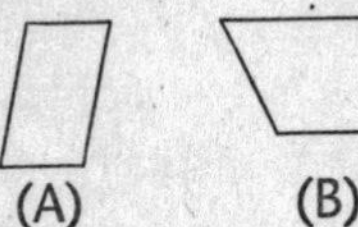
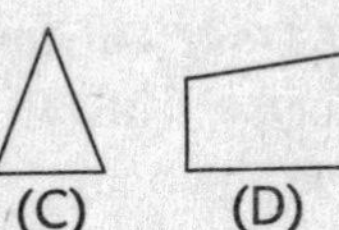

(A) (B) (C) (D)

Parentheses *Used to enclose portions of mathematical expressions; due to the parentheses in the expression $(7 - 3)^2$, 3 must be subtracted from 7 before squaring the quantity.*
When it comes to her friendship with Gabby and Wendy, Sonia sometimes feels as though the other girls are inside a set of **parentheses** while she's on the outside.

Simplify $5 \times (3 + 1)$.

Pattern *A predictable sequence; the sequence 0, 2, 4, 6, 8, 10,... is a pattern of even numbers; the sequence 3, 6, 9, 12, 15, 18,... is a pattern of multiples of 3.*

Due to a slight mishap with a beaker of black-colored liquid in the Chemistry Corner, Anna now had a cool, splotchy **pattern** on her favorite T-shirt.

Write the missing number: 1, 2, 4, 8, 16, ___

Pentagon *A 5-sided polygon.*

Mr. Stickler sat in the café, chewing his gum and contemplating the **Pentagon** in Washington D.C. Was the famous building named that way because it was a five-sided polygon?

Which of the following figures is a pentagon?

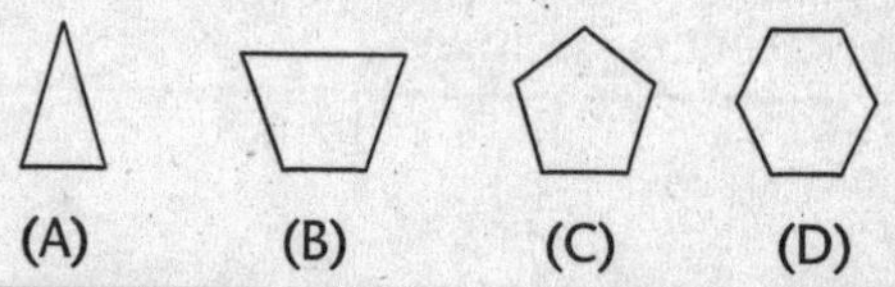

Percent (%) *A number out of 100; 25% means* $\frac{25}{100}$.

Anna was 75% sure that Mr. Stickler was going to have the class write a report on the science museum.

Convert 0.6 to a %.

Percent of decrease *The ratio of the amount of decrease to the original amount expressed as a percent.*

Because Anna was 75% sure about getting a museum assignment, her **percent of decrease** was 25% of being totally sure.

Connor's Coffee Stand sold 800 coffees during February and 600 coffees during March. Find the percent of decrease.

Percent of increase *The ratio of the amount of increase to the original amount expressed as a percent.*

Mr. Stickler was surprised to find that biscotti now cost $2 each as opposed to $1 each just an hour before. That was a one hundred **percent of increase!**

Cindy's Ice Cream Shop sold 500 milkshakes in May and 700 milkshakes in June. Find the percent of increase.

Perfect square *A number that is the square of an integer; 36 is a perfect square because $36 = 6^2$.*

Wendy pointed out the identical twins girls at the exhibit next to her. They were the **perfect square!**

List the first 10 perfect squares.

Perimeter *The distance around a closed figure.*
Anna walked the **perimeter** of the 3-D theater before she found the seat that she wanted to sit in.

Find the perimeter of the rectangle below.

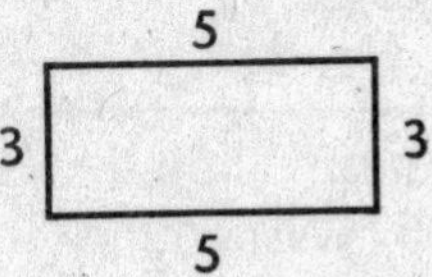

Perpendicular *Lines or line segments that form right angles when they intersect.*
When Mr. Stickler spotted Chris and Lester arguing, he placed his hands in a **perpendicular** position making his infamous "T"-for-time-out signal.

Which 2 lines in the diagram below are perpendicular?

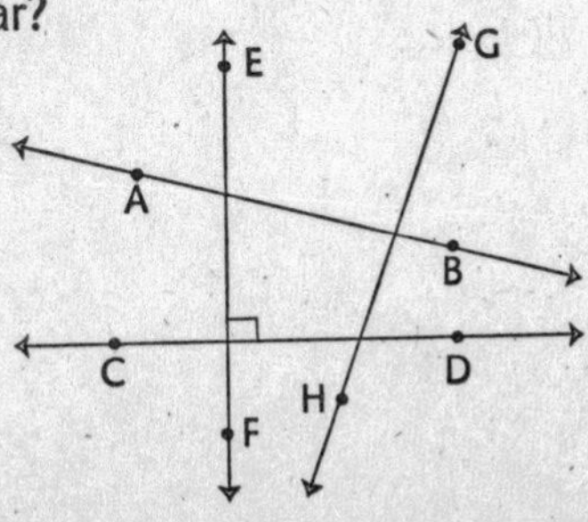

Pi *The ratio of the circumference of a circle to the diameter of the circle (3.141592...); represented by the symbol π and often approximated as 3.14 or $\frac{22}{7}$.*

Chris rolled his eyes when Lester handed him a slip of paper with the symbol π on it. Chris had remarked that he wanted pie—*not* **pi!**

What is the value of the circumference of a circle if the diameter equals 5?

Pint *A customary unit of capacity; 1 pint = 2 cups.*

Thrilled that Wendy, Sonia, and Gabby admired her black-splotched tee, Anna gladly splashed a **pint** of black liquid onto their shirts too.

Fill in the blank: 4 pints = _____ cups.

Place value ***The value given to a digit due to its position relative to the decimal point's place.***
For example: In the number 635.8, the 6 is in the hundreds ***place,*** *and its value is 600; the 8 is in the tenths place, and its value is* $\frac{8}{10}$.

It's not as if Johnny wasn't grateful for finding 15 dollars in his pocket. It's just that he wished he could switch the **place values** of the 1 and the 5.

In the number 452, what is the value of the 5?

Plane ***A flat surface that continues in all directions; 3 points not contained in the same line define a plane.***
Lester's Superball accidentally slipped out of his pocket and bounced all around the plane of the Rockin' Robots activity table.

Name the plane shown below.

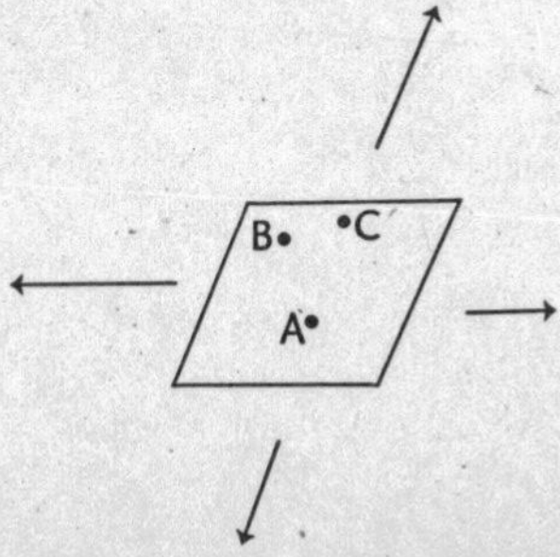

Plot *To locate a point on a map or coordinate plane.*
Anna used her museum guide map to **plot** the next exhibit she'd visit.

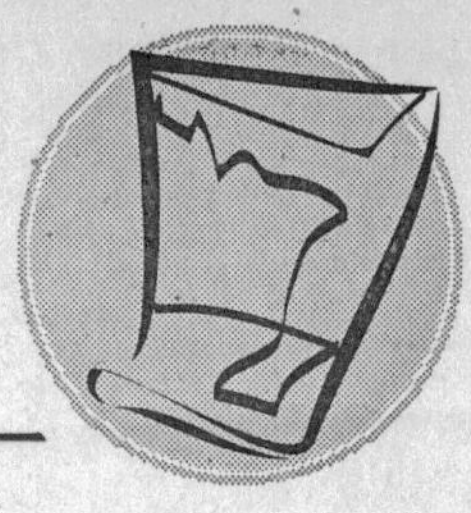

Plot the point A (–2, 4).

Point *A location in space.*
When Wendy asked Sonia, "What's your **point?**" she wasn't looking for an explanation. She was looking for a location.

Which point is located on circle A?

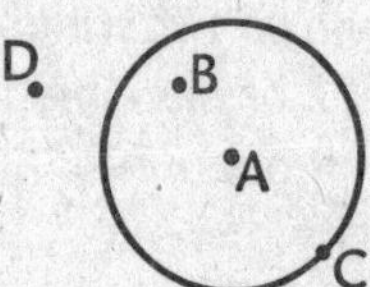

Polygon *A plane, closed figure with 3 or more sides.*

Now that Anna was part of the group, Wendy, Sonia, and Gabby revised their super-secret greeting of touching index fingers to form a triangle to making a **polygon** with their ring fingers.

Which of the following figures is a polygon?

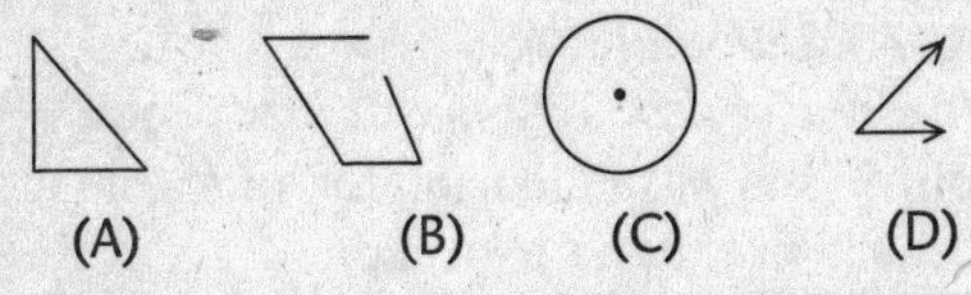

(A) (B) (C) (D)

Polynomial *An expression containing a monomial or the sum of 2 or more monomials.*

Perhaps if Lester hung around Anna, Wendy, Sonia, and Gabby long enough, they'd let him join their **polynomial** too . . . or perhaps not.

Can a polynomial contain more than 3 terms?

Positive *A number that is greater than 0.*
It was hard to believe that one stained T-shirt could transform Anna's social life and give her a **positive** number of friends, as opposed to 0.

Place 3, – 4, – 2, and 5 on the number line below.

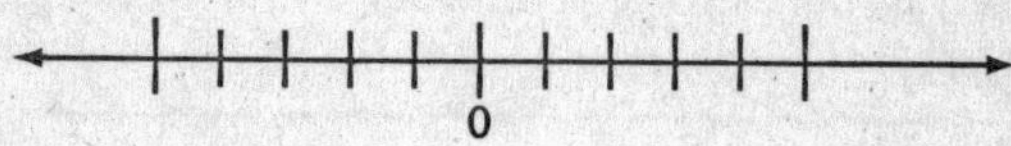

Pound *A customary unit of measure; 1 pound = 16 ounces.*
Mr. Stickler stepped on the scale in the Weights and Measures center and found that he had gained three **pounds**! Could it have been the biscotti?

Fill in the blank: 1 pound = _____ cups.

Power *An expression written as x^n is read as x to the nth power; 3^4 is read as 3 to the 4th power; $3^4 = 3 \times 3 \times 3 \times 3 = 81$.*
Anna felt like a "girl to fourth **power**" as she and her three new friends marched through the museum with identical T-shirts.

What happens to a number when you raise it to the first power?

Prime number *A number with exactly 2 factors, 1 and itself; 0 and 1 are neither prime nor composite; the first 5 prime numbers are 2, 3, 5, 7, and 11.*
Lester told Anna that they were "in their prime" at 11 years old—their **prime number,** that is.

Which of the following numbers is prime?
(A) 5 (B) 9 (C) 20 (D) 21

Prime factorization *Expressing a composite number as a product of prime numbers; 5 × 5 × 3 is the prime factorization of 75.*
Lester also mentioned that 11 × 11 is the **prime factorization** of 121.

Find the prime factorization of 100.

Prism *A three-dimensional figure whose bases are parallel, congruent polygons and whose sides are parallelograms*
Anna admired the cute little crystal **prisms** dangling from Sonia's ears.

Which of the following figures is not a prism?

(A)
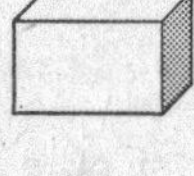
(B)
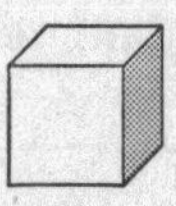
(C)

(D)

Probability *The chances that an event will occur; the ratio of the number of desired outcomes to the number of all possible outcomes.*

$$P(E) = \frac{\text{number of desired outcomes}}{\text{total number of possible outcomes}}$$

Anna quickly figured that the **probability** of Sonia giving Anna her cool earrings was not that probable.

What is the probability of getting a 5 when you roll a die?

Product *The result of multiplication.*
In front of the Wonders of the Eye exhibit, Wendy admitted that her baby blues were the **product** of good genes.

Find the product of 4 and 7.

Proper fraction *A fraction with the denominator larger than the numerator; the value of a proper fraction is always less than 1;* $\frac{2}{7}$ *is a proper fraction.*

Anna would never admit that she was psyched to be $\frac{1}{4}$ of a cool group of girls because it happened to be a **proper fraction,** but it was true.

Which of the following is a proper fraction?

(A) $\frac{5}{4}$ (B) $\frac{7}{2}$ (C) $\frac{3}{5}$ (D) $\frac{9}{6}$

Proportion *Two equal ratios.*

Wendy asked Sonia if the amount of eyeliner she was wearing was in **proportion** to the lip gloss she had on.

What must *n* equal to make the following proportion true?

$$\frac{2}{5} = \frac{2 \times n}{20}$$

Protractor *An instrument used to measure angles.* Before Chris had visited the Geometry Genius exhibit, he'd only pretended to know how to measure angles with a **protractor.** Now he really knew how to do it!

Using a protractor, measure the angle below.

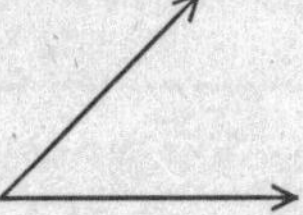

Pyramid *A three-dimensional shape whose base is a polygon and whose sides are triangles that meet at a point.* Anna snapped a picture of Wendy, Sonia, and Gabby forming a human **pyramid** in front of the Ancient Egypt exhibit.

Find the total number of edges for a pyramid with a square base.

Pythagorean Theorem *In a right triangle, if a and b are the legs and c is the hypotenuse, then* $a^2 + b^2 = c^2$.

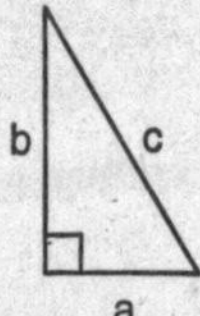

Chris used his protractor to make sure he was working with a right angle before he tested the validity of the **Pythagorean Theorem:** $a^2 + b^2 = c^2$.

Can the Pythagorean Theorem be used to find the missing side in the triangle below?

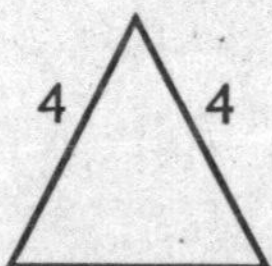

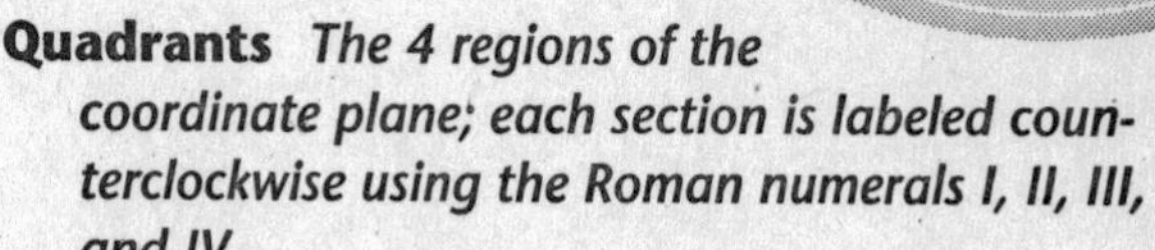

Quadrants *The 4 regions of the coordinate plane; each section is labeled counterclockwise using the Roman numerals I, II, III, and IV.*

Johnny didn't get it. So what if his sneeze landed across all 4 **quadrants** of the table? It's just a little booger.

In what quadrant does the point A (5, –1) lie?

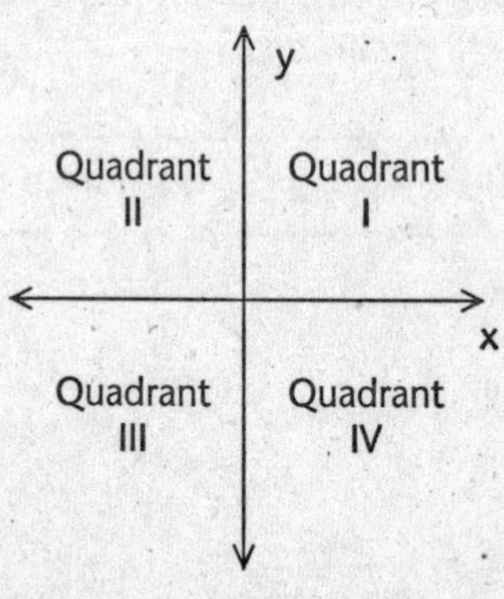

Quadrilateral ***A four-sided polygon; the sum of the interior angles is 360°.***

Johnny liked the odd shape of the eco-house because it was round, unlike the boring **quadrilateral**-shaped buildings around it.

Which of the following figures is a quadrilateral?

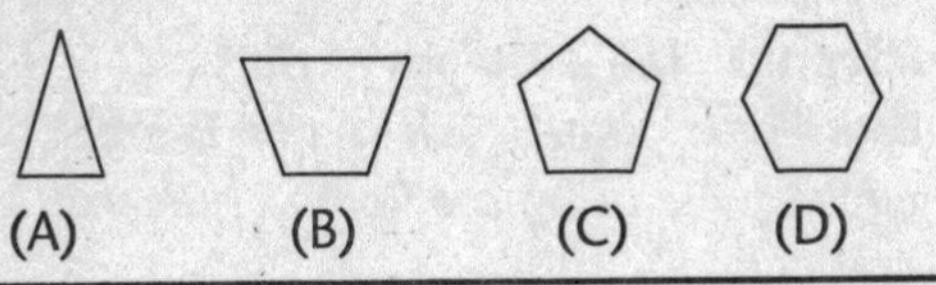

Quart ***A customary unit for measuring capacity; 1 quart = 2 pints = 4 cups.***

Mr. Stickler thought the **quart** of muddy water in the Secrets of the Swamp exhibit reminded him of chocolate milk.

Fill in the blank ______ quarts = 8 cups.

Quotient ***The result of a division problem; dividend ÷ divisor = quotient.***
Wendy divided her chocolate bar by 4 and handed out the quotient to her friends.

Which number in the equation 30 ÷ 5 = 6 is the quotient?

Radical *A symbol that indicates a root; the symbol $\sqrt{\ }$ means square root; the symbol ($\sqrt[3]{\ }$) means cube root.*
Although Lester thought it hysterical to tell Chris that solving the square root of 25 was **radical,** Chris did not laugh.

Find the value of $\sqrt{16}$.

Radicand *The number or quantity contained under the radical sign.*
Anna felt a bit like a **radicand** as she stood underneath the sweeping banana-tree leaves in the Horticulture Hut.

In the expression, $\sqrt{3} - 7 + 52$, which number is the radicand?
(A) 2 (B) 3 (C) 5 (D) 7

Radius *A line segment whose endpoint are the center of a circle and any point on the circle; the diameter equals twice the length of the radius.* Johnny guessed that the architect who designed the round eco-house had to, at some point, figure out its **radius.**

Which segment in the diagram below is a radius of circle F?

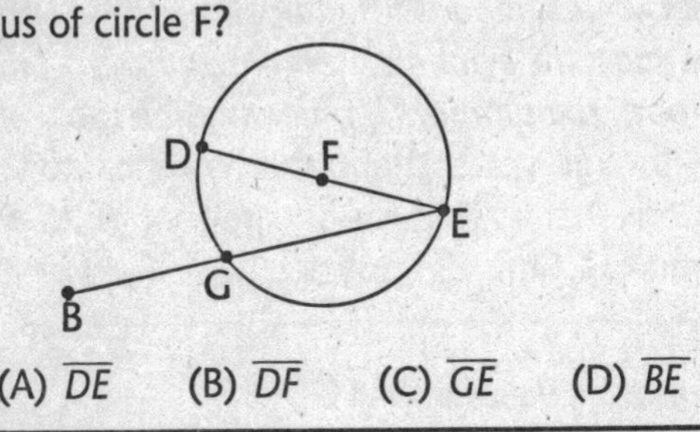

(A) $\overline{DE}$ (B) $\overline{DF}$ (C) $\overline{GE}$ (D) $\overline{BE}$

Raised to *To multiply a number by itself the given number of times; 5 raised to the 4th power means $5^4 = 5 \times 5 \times 5 \times 5 = 625$.* Johnny vowed to tell his parents about the eco-house. If the family lived in one they'd save not just 10 bucks a year, but ten bucks **raised to** the third power: $1,000!

What is the value of the expression "2 raised to the 3rd power"?

Range ***The highest value minus the lowest value in a set of data.***
If Johnny's parents saved $1,000 a year, living in an eco-house, that meant he could ask them to buy him a new bike that cost anywhere in the **range** of $1 and $1,000.

A student's first 5 test scores were 80, 91, 97, 88, and 83. What is the range of these scores?

Rate ***A ratio comparing different units; often uses the key word "per."***
Lester noted that his "museum buddy" Chris had tried, and failed, to ditch him at a **rate** of 2 times per exhibit.

If Anna travels at a constant speed of 30 miles per hour for 2 hours, how far has she traveled? (Hint: Use the formula *distance* = *rate* × *time.)*

Ratio *A fraction that compares 2 quantities; can be expressed as a:b, a to b, or* $\frac{a}{b}$.

Mr. Stickler couldn't believe his eyes when he read the wrapper of his chocolate bar. The "Win a Free Trip to Vegas" sweepstakes he'd just won had odds with a **ratio** of 250,467,039 to 1!

Which shows the ratio 2:3?

○○○ □□□	○○○○ □□□□	○○○○ □□□□□□	○○ □□□□
(A)	(B)	(C)	(D)

Rational *A real number that can be expressed as the ratio of 2 integers (denominator not 0) or as terminating or repeating decimals.*

Anna just realized that today she was exactly 11½ years old, which happened to be a **rational** number.

Which of the following numbers is rational?

(A) 7.120120012 . . . (C) $6.\overline{2}$

(B) $\sqrt{5}$ (D) π

Ray ***Part of a line that has an endpoint and extends infinitely in one direction; ray AB is written as $\overline{AB}$.***
Wendy's eyes looked as if they were shooting death **rays** at the tall girl with the big head, who'd sat in front of her at the 3-D theater.

Which of the following could not be used to name the ray in the diagram below?

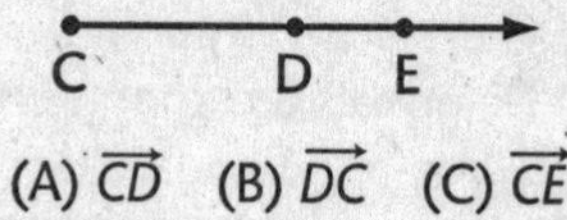

(A) $\overrightarrow{CD}$ (B) $\overrightarrow{DC}$ (C) $\overrightarrow{CE}$

Real numbers ***All of the points on the number line.***
Mysteries of the Ocean, the 3-D movie, taught Anna that the amount of species living in the sea were as infinite as a set of **real numbers.**

Use your estimation skills to place the following real numbers on the number line below.

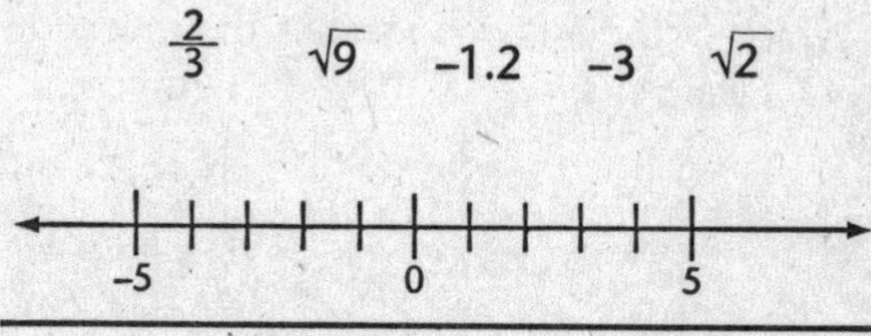

Rectangle ***A parallelogram with 4 right angles; Perimeter = 2l + 2w; Area = bh.***

Unfortunately Wendy could not see *Mysteries of the Ocean* because her view was blocked by a girl with hair shaped like a **rectangle.**

Which of the following is not a property of all rectangles?

(A) Opposite sides are equal.
(B) There are 4 right angles.
(C) Diagonals are congruent.
(D) There are 4 equal sides.

Reflection ***A transformation that creates the mirror image of a figure by "flipping" it over a line or point; the symbol $r_{x\text{-}axis}$ denotes a reflection over the x-axis.***

A little bored, Wendy whipped out a tiny mirror and stared at her **reflection** in it.

Which of the following shows the reflection of $\overline{AB}$ over line k?

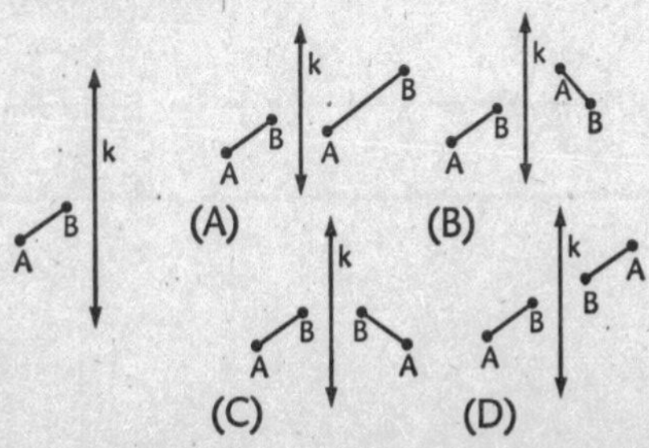

Remainder *The amount left after dividing one number by another; 7 ÷ 3 = 2 with a remainder of 1.*

Wendy was horrified to find a **remainder** of her spinach salad in between her 2 front teeth!

Find the remainder when 17 ÷ 5.

Repeating decimal *A decimal whose digits infinitely repeat in groups of one or more; $4.3\overline{1}$ means 4.311111...*

No matter how carefully Wendy tried not to lodge food in her teeth, history kept repeating itself over and over and over, making her feel like a **repeating decimal** with infinite digits.

Write 72.151515151515 . . . using the repeating symbol.

Rhombus *A parallelogram with 4 congruent sides.*

Wendy was just about to get the last of the spinach, when she accidentally dropped her **rhombus**-like mirror onto the floor.

Which of the following cannot be classified as a rhombus?

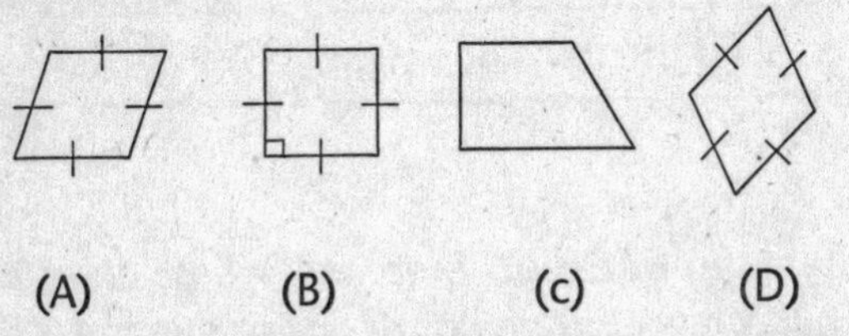

(A) (B) (c) (D)

Right angle *An angle that measures 90°.*

When she retrieved her rectangle-shaped mirror, Wendy found that a diagonal crack had formed between 2 of the **right angles.**

Which of the following is a right angle?

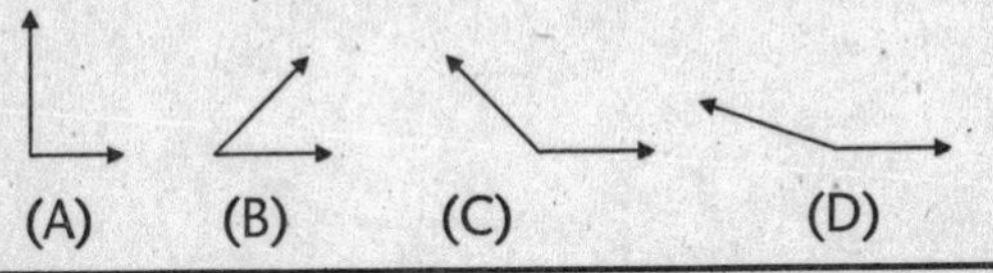

(A) (B) (C) (D)

Right triangle *A triangle that contains 1 right angle.*

The diagonal crack had turned Wendy's simple rectangle-shaped mirror into 2 **right triangles.**

Which of the following represents the measures of the angles of a right triangle?

(A) 44°, 46°, and 90°
(B) 30°, 50°, and 100°
(C) 41°, 60°, and 79°
(D) 50°, 60°, and 70°

Rotation *A transformation that turns a figure a certain number of degrees around a fixed point or line.*

Wendy gave her cracked mirror a 360° **rotation** to check it out.

In the diagram below, which transformation maps ΔABC to $\Delta A'B'C'$?

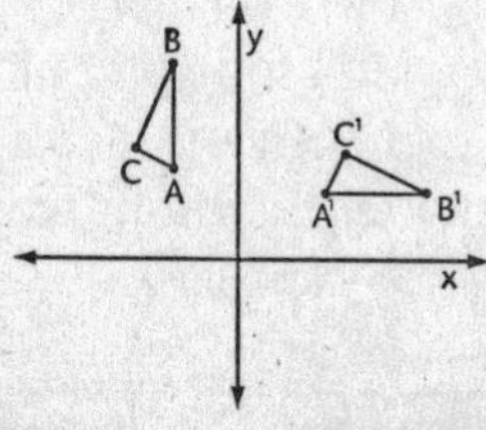

(A) Dilation
(B) Reflection
(C) Rotation
(D) Translation

Rotational symmetry ***When a figure can be turned less than 360° around a fixed point and look exactly like it did before it was turned, it has rotational symmetry.***

When Wendy turned her cracked mirror 180°, she realized that it had **rotational symmetry** because it looked exactly the same as it had before she turned it.

Which of the following letters has rotational symmetry?

(A) A (B) G (C) C (D) S

Round ***To approximate the value of a number to a given decimal place; if the digit to the right of the given decimal place is greater than or equal to 5, round up.***

For Example: 36.47 rounded to the nearest tenth is 36.5 because the digit to the right of the tenths place (the hundredths place) is a 7; 36.42 ***rounded*** *to the nearest tenth is 36.4 because the digit to the right of the tenths place is 2.*

Wendy wasn't too upset about the broken mirror. She told everyone it cost her a dollar, but that was after she **rounded** it up from its real price: ninety-nine cents.

Round 46.82 to the nearest tenth.

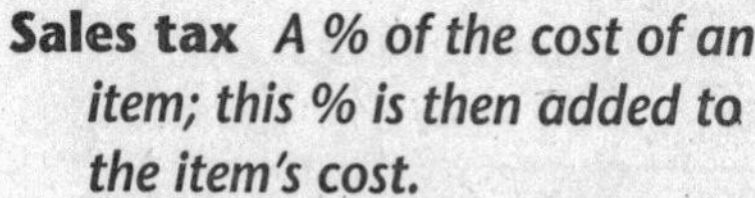

Sales tax *A % of the cost of an item; this % is then added to the item's cost.*
Mr. Bickman thought that there ought to be a law against paying **sales tax** on chocolate because it always made his snacks cost just a little bit more.

What is the cost of a $20 shirt after an 8% sales tax is added?

Scalene triangle *A triangle with no equal sides.*
Much to Chris's embarrassment, Lester showed off his geometry talent at the Geometry Genius exhibit by using 3 different sized fingers to form a **scalene triangle** with his hands.

Which of the following triangles is not scalene?

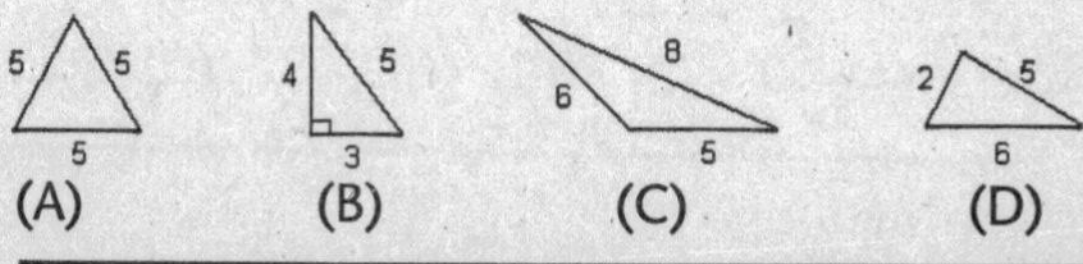

Scientific notation *A system that expresses a number as a number between 1 and 10 multiplied by a power of 10; 320,000 expressed in scientific notation equals 3.2×10^5; 0.0049 expressed in scientific notation equals 4.9×10^3.*
Chris thought expressing Lester's dweeb factor in **scientific notation** might help Lester understand how nerdy he was acting. On a scale of 1 to 500, Chris told Lester he was a 5×10^2.

Write 491,000,000 in scientific notation.

Semicircle *Half of a circle whose boundaries are formed by an arc and a diameter.*
Johnny joined the **semicircle** that had formed in front of a display of dancing skeletons in the human body gallery.

In the diagram below, which of the following arcs is a semicircle?

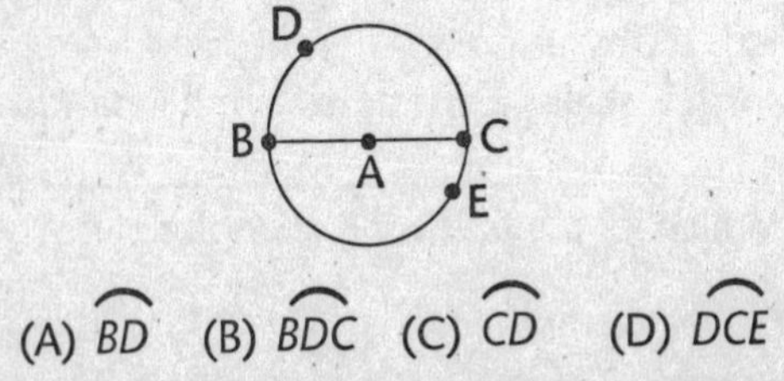

(A) $\overset{\frown}{BD}$ (B) $\overset{\frown}{BDC}$ (C) $\overset{\frown}{CD}$ (D) $\overset{\frown}{DCE}$

Set ***A grouping of items that have something in common.***
Anna, Gabby, Wendy, and Sonia became known to the museum employees as "the **set** of girls, wearing the black-splotched T-shirts."

Which of the following represents the set of odd numbers?

(A) {2, 4, 6, 8,...} (C) {3, 6, 9, 12,...}
(B) (1, 3, 5, 7,...} (D) {4, 8, 12, 16,...}

Similar ***Same shape, not necessarily the same size; corresponding angles are congruent; corresponding sides are proportional.***
The black splotch on each girl's T-shirt was **similar** but not identical.

Which triangle is similar to  ?

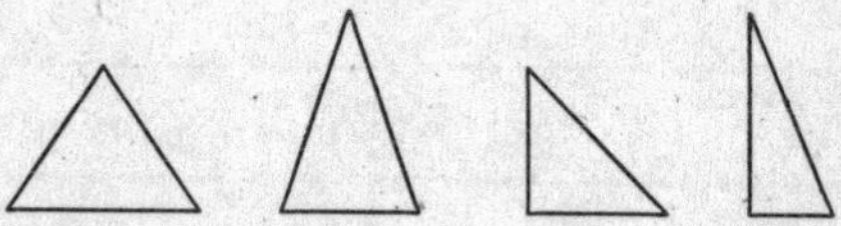

(A) (B) (C) (D)

Simple interest *Interest calculated by multiplying the principal times the interest rate times the time in years; formula is* $I = p \times r \times t$.

For Example: A principal of $1,000 is borrowed for 2 years at 8%. Find the interest.

$$I = p \times r \times t$$
$$I = 1{,}000 \times .08 \times 2 = \$160$$

Lester agreed to lend Chris $10 as long as Chris paid the **simple interest** of $125 by the end of the week.

A principal of $2,000 is borrowed for 5 years at 7%. Find the interest.

Simplify *To perform operations in an expression to make a problem less complicated.*

Sonia decided to **simplify** her outfit by removing her cabbie hat, 6 bangle bracelets, and a pair of legwarmers, leaving her in jeans and a T-shirt.

Simplify $5^2 - 4 \times 6$.

Slope *The ratio of the change in y-coordinates (rise) to the change in x-coordinates (run); describes the steepness of a line; m is the symbol used for slope in the equation y = mx + b.*
Gabby miscalculated the **slope** of the ramp leading to the caveman exhibit and fell flat on her butt.

Which of the following lines in the graphs below has a slope of $\frac{2}{3}$?

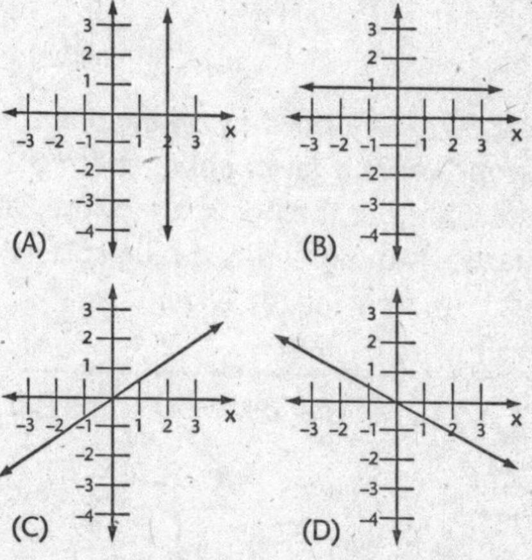

Solid *A three-dimensional shape.*
"Excuse me," Chris said after he accidentally bumped into a **solid** teacher in the Hall of Holograms.

Which solid is a cylinder?

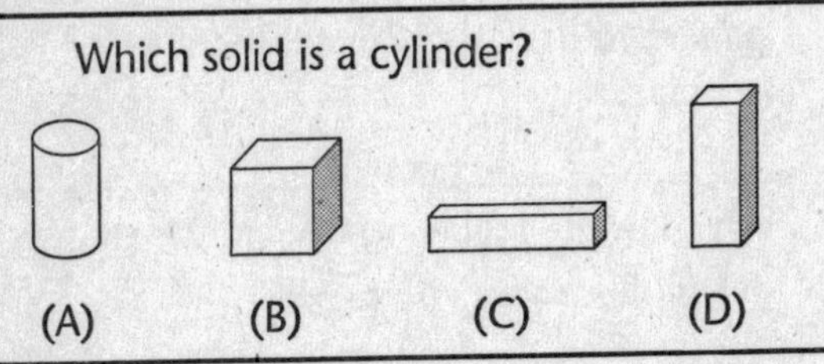

(A) (B) (C) (D)

Sphere *A solid formed by all the points a given distance from a given point, called the center.*
Chris stared at the hologram of the Earth, amazed as the green-and-blue **sphere** spun past him, only inches away.

Which of the following solids is a sphere?

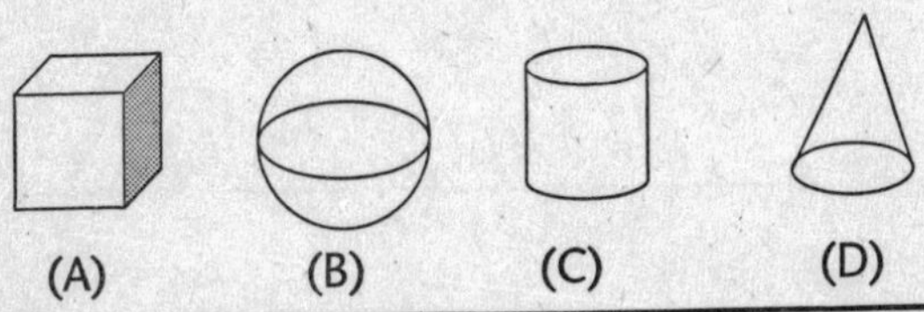

(A) (B) (C) (D)

Square ***A parallelogram with 4 congruent sides and 4 right angles; perimeter = 4s and area = s^2, where s equals the length of 1 side.***
When Mr. Stickler was young, the kids used to call him a "**square**" because they thought he was as dull and predictable as a parallelogram with 4 congruent sides and 4 right angles.

If one side of a square equals 5, find the perimeter.

Squared ***Multiply a number by itself; x^2, read as x squared, means $x \times x$.***
Sometimes Wendy wished that she were **squared** so that Gabby and Sonia wouldn't fight over who got to sit next to her all the time.

What is the value of 7 squared?

Stem-and-leaf plot *A way to arrange data using place value.*

Mr. Stickler decided that when they got back to school on Monday, he would have his class rate the field trip from 1 to 100, and he would then show them how to compile all the data into a **stem-and-leaf plot**—it would be a great way to teach a lesson!

The top golf scores of a golf team are as follows: 85, 93, 81, 79, 98, 78, 93, 85, 93, 79. Create a stem-and-leaf plot to organize the data.

Straight angle *An angle that measures 180°; a straight line.*

Tired, Johnny decided to make like a **straight angle** and take a nap in the museum's planetarium.

Which of the following angles is a straight angle?

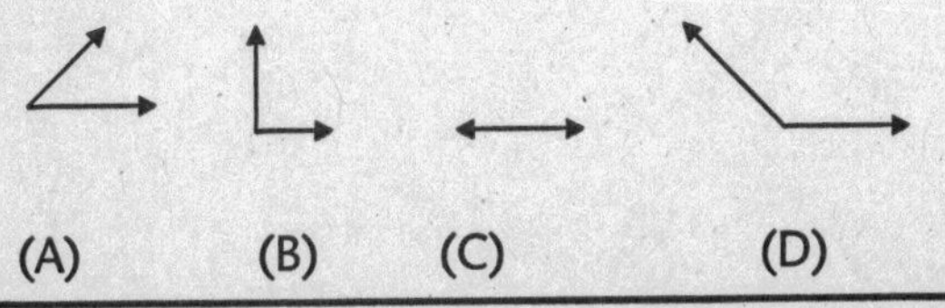

(A) (B) (C) (D)

Subtract *To take one amount away from another.*

Lester could imagine Chris as a cool dude only if he **subtracted** the guy's bad attitude.

Subtract 10 from 25.

Substitute *To replace a variable with a number.*

No way was the row of hard plastic seats in the planetarium a **substitute** for Johnny's soft comfortable bed at home.

Use substitution to check if $x = 6$ in the equation $x + 4 = 10$.

Sum (+) *The result of adding 2 numbers or quantities.*

Once in the planetarium, Anna counted the number of known planets in our solar system and came to the **sum** of 9.

Find the sum of 25 and 15.

Supplementary angles *Two angles whose measures have a sum of 180°.*
Sonia and Gabby made **supplementary angles** when Sonia leaned back in her seat, making a 160° angle and Gabby leaned forward, forming a 20° angle.

If the measure of an angle is 80°, find the measure of its supplement.

Surface area *The sum of the areas of all of the faces on a three-dimensional figure.*
Anna noticed that the **surface area** of the planetarium floor was dotted with chewed up wads of gum.

Find the surface area of a cube whose edges equal 3.

T

Tens place *The place value that is 2 spaces to the left of the decimal point.*
Gabby, Anna, Wendy, and Sonia giggled when they heard Johnny sleeping in the planetarium and rated his snores in the **tens.**

Which digit is in the tens place in the number 598.491?

Tenths place *The place value one space to the right of the decimal point.*
Gabby was disappointed when only **two-tenths** into the laser light show at the planetarium, the laser broke down.

Round 321.48 to the nearest tenth.

Terminating decimal *A number that has a finite amount of digits after the decimal point.* Anna sure was glad that the price of her soda, \$1.25, had a **terminating decimal** because that was all the money she had left.

Which of the following is a terminating decimal?

(A) 4.75 (C) $4.7\overline{5}$

(B) $4.\overline{75}$ (D) 4.75075007500075...

Term *The product of numbers and variables; the expression $3x^2 - 5ab + 4$ contains the following 3 terms: $3x^2$, $5ab$, and 4.* Mr. Stickler told the kids that in no uncertain **terms** would they get out of writing a report on their trip to the museum.

How many terms are in the expression $2x + 3y - 7z + 8$?

Thousandths place ***The place value that is 3 spaces to the right of the decimal point.*** Lester couldn't figure out why Chris didn't like him. Maybe it was because Chris had a brain capacity that was two-**thousandths** the size of Lester's.

Which digit is in the thousandths place in the number 9,346.2875?

Three-dimensional ***A figure that has 3 measurable qualities: length, width, and height.*** The picture of the Underwater Adventure exhibit was so cool that Johnny couldn't wait to see the **three-dimensional** version.

Which of the following figures is three-dimensional?

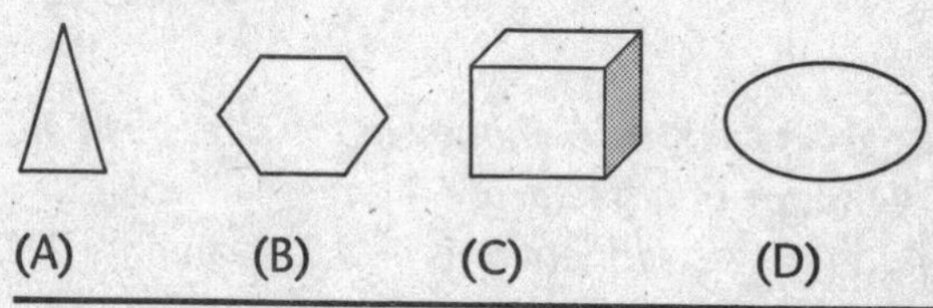

Times *Multiply; symbol (× or ·).*
Everyone saw the Underwater Adventure exhibit only once, but Johnny found it so interesting that he went back to look at it 3 **times** more than everyone, which meant that he saw it a total of 3 times.

Find the value of 3 times 7.

Ton *A customary unit of measure; 1 ton = 2,000 pounds.*
Mr. Stickler always knew his hybrid car was his "**ton** of fun," but the Future of Fuel exhibit showed him that he was helping to save the environment by driving it.

Fill in the blank 3 tons = _____ pounds.

Transformation *A change in the size, position, or shape of a geometric figure.*
Johnny was saddened to find that global warming had caused the **transformation** of some coral reefs, once pink and lush, to gray and lifeless.

Which transformation slides a figure to a new location?

(A) Dilation (C) Rotation
(B) Reflection (D) Translation

Translation ***A transformation that slides a figure, without flipping or turning it, to a new location.***

When Gabby got up to use the bathroom, Sonia slid over into her seat, resulting in a **translation** of one seat to the right and one row forward.

In which of the following pairs is one figure a translation of the other?

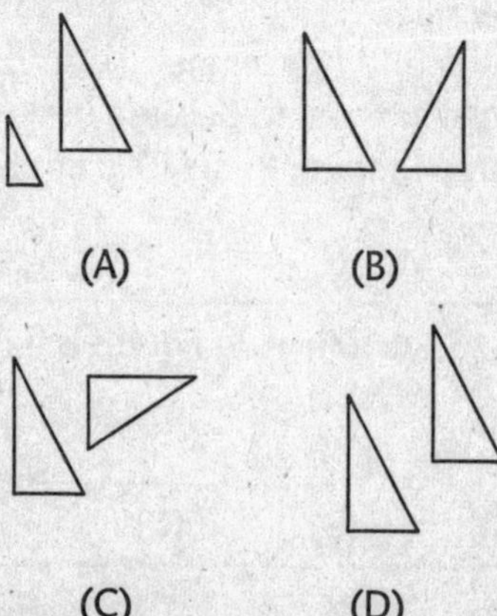

Transversal *A line that intersects at least 2 other lines.*
Anna realized that she is a **transversal** student because she can hang out with the popular girls as well as with the bookish crowd.

Which line in the diagram below is a transversal?

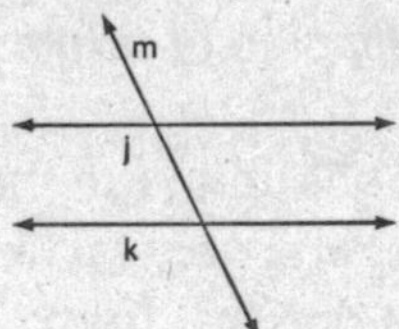

Trapezoid *A quadrilateral with exactly 1 pair of parallel sides.*
Chris thought the Geometry Genus room was especially interesting because it was shaped like a **trapezoid** with only 1 pair of parallel sides.

Which of the following figures is a trapezoid?

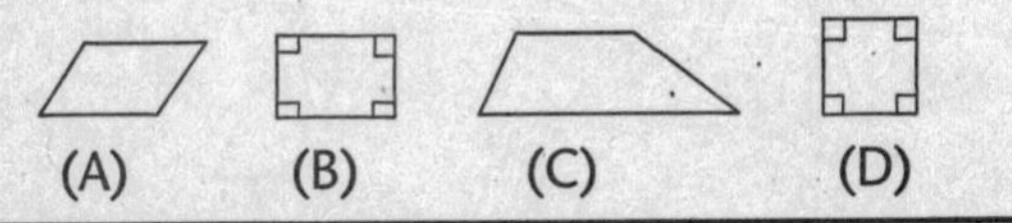

Tree diagram *A diagram that shows all of the possible outcomes of an event.*

Having a crush on Anna, Lester drew a **tree diagram** to see if he should ask her to go to the Underwater Adventure exhibit with him . . . although there were more negative than positive outcomes, he asked her anyway.

A coin is tossed twice. Draw a tree diagram to display the possible outcomes.

Triangle *A 3-sided polygon; the sum of the interior angles is 180°.*

Anna fiddled with her **triangle**-shaped earrings, trying to decide if she wanted to hang out with Lester.

Classify this triangle by its angles and sides.

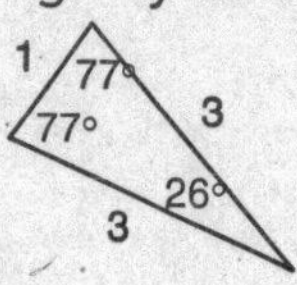

Trinomial *A polynomial with 3 terms.*

Anna agreed to hang out with Lester only if he brought along 1 more person, making their polynomial a **trinomial.** Lester said no problem and told her that Chris would tag along.

Which of the following expressions is a trinomial?

(A) $-9x$ (C) $y - 105$
(B) 321 (D) $3x + 2y - z$

Two-dimensional *A figure that has 2 measurable qualities: length and width.*

Anna admitted that the **two-dimensional** photo of the Underwater Adventure exhibit did not do the display justice.

Which of the following figures is two-dimensional?

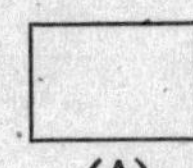
(A)

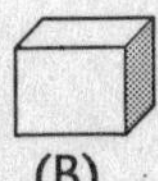
(B)

(C)

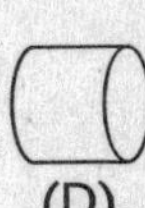
(D)

V

Variable *A letter that represents a number or set of numbers in an expression or equation.*

Lester had to admit that with all the **variables** in his tree diagram, chilling with Anna seemed the least likely to have happened.

Which of the following is the variable in the term $-8y^2$?

(A) –8 (C) 8

(B) 2 (D) *y*

Venn Diagram *A diagram that uses circles to show the relationship between sets of data; has one or more loops.*

Lester quickly penciled a few circles, creating a **Venn Diagram** that plotted the course of his relationship with Anna—from nonexistent to new friendship.

In the Venn Diagram below, 1 loop represents odd numbers and the other loop represents numbers greater than 10. Notice how numbers that match both attributes go in both loops.

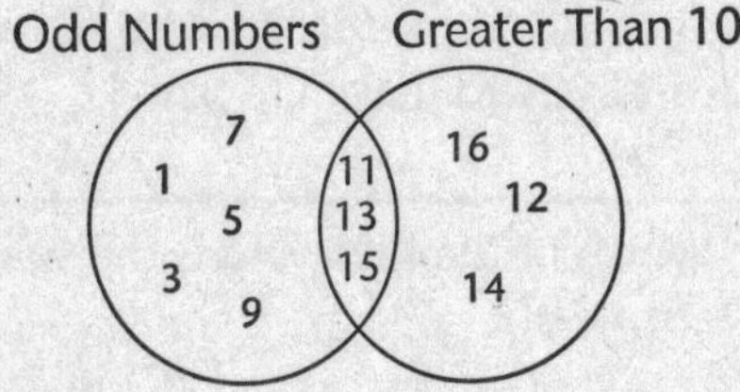

Draw a Venn Diagram with 1 loop where all the numbers inside are divisible by 2 and with another loop where all the numbers inside are divisible by 3. Put at least 12 numbers in the diagram. Three of the numbers should be in both loops.

Vertex ***The common endpoint of 2 rays; the plural of vertex is vertices.***
Lester accidentally bumped into Chris at the **vertex** to their path to the exhibit entrance.

Which of the following is the vertex of $\angle ABC$?

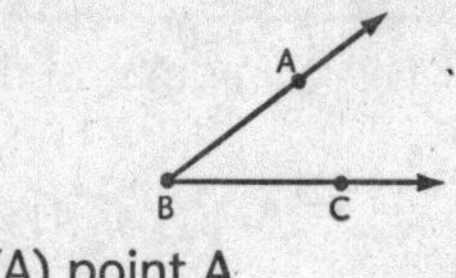

(A) point A
(B) point B
(C) point C
(D) $\overline{AB}$

Vertical ***A line going straight up and down; a wall is vertical.***
Waking up from his nap, Johnny got **vertical** just in time to catch the bus back to school.

In the graph below, which axis is vertical?

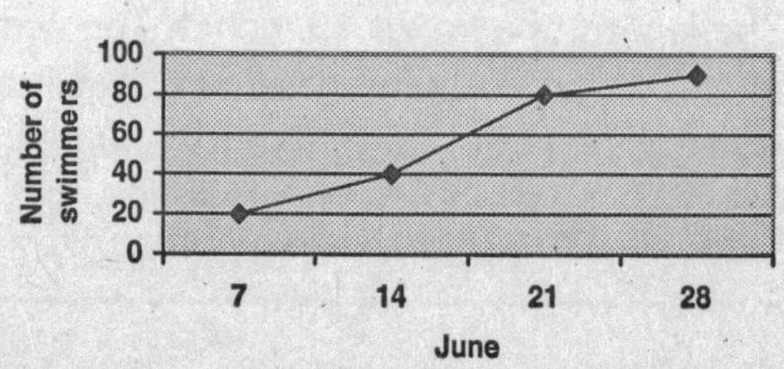

Vertical angles ***The opposite angles formed when 2 lines intersect.***

Anna commented on how the last 2 pizza slices in the museum's cafeteria looked like cheesy **vertical angles** because their tips were touching.

Name a pair of vertical angles in the diagram below.

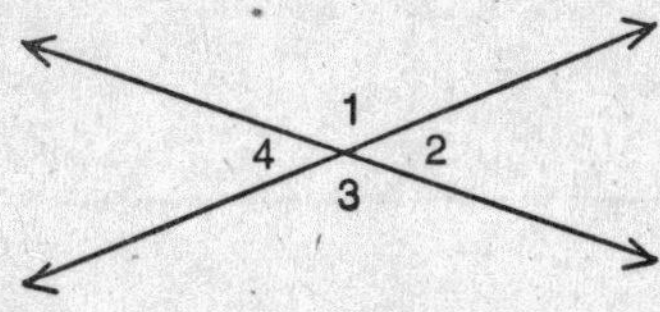

Volume ***The number of cubic units needed to fill the space inside a solid.***

Before meeting the rest of the class to go home, Anna ate a final slice of pizza that filled the entire **volume** of her stomach.

The length of a box is 10 inches. The width is 5 inches. What other information about the box do you need in order to find its volume?

(A) The weight (C) The height
(B) The contents (D) The circumference

W

Weight *Heaviness; determined by the mass of an object and the force of gravity.*

Maybe Anna shouldn't have eaten the pizza because now she felt as if she had swallowed a 10-pound **weight.**

In the customary measurement system, which of the following units of measurement is not used when measuring weight?

(A) ounces (C) pounds
(B) gallons (D) tons

Whole numbers *The set of numbers {0, 1, 2, . . .}.*
Mr. Bickman counted to 100, using **whole numbers,** as he waited impatiently for his class to meet him by the front doors of the museum. He hoped that nobody would be late.

Which of the following numbers is not a member of the set of whole numbers?

(A) –2 (B) 0 (C) 3 (D) 1,204

Width *One dimension in a two-dimensional or three-dimensional shape.*
Johnny ran half the **width** of lobby, then went straight for the front doors, and reached Mr. Bickman first.

Find the area of a rectangle below:

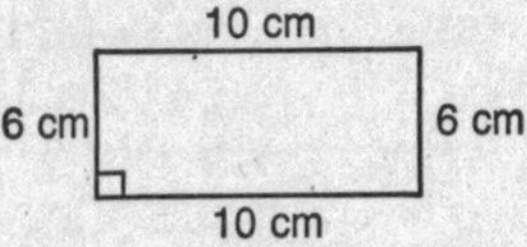

X-Z

x-axis *The horizontal number line on the coordinate plane.*
Wendy made like an **x-axis** and raced across the lobby with Sonia and Gabby following close behind.

Is the x-axis vertical or horizontal?

---

x-intercept *The point where a line intersects the x-axis.*

Avoiding a lady who was crossing vertically through the lobby, Gabby stopped suddenly right before the **x-intercept.** Unfortunately Sonia bumped into her, knocking them both to the ground.

What is the *x*-intercept of line *m*?

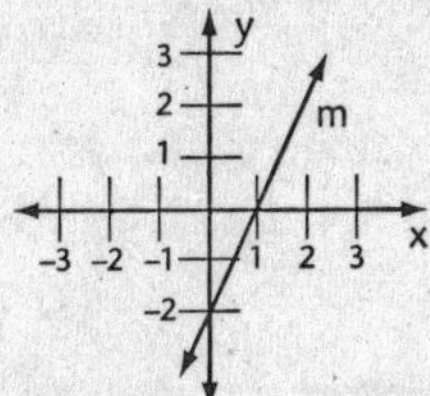

Yard *A customary unit of length; 1 yard = 3 feet = 36 inches.*

Gabby and Sonia laughed because they'd almost made it to where the rest of the class was standing, just one **yard** away.

Fill in the blank: 4 yards = _____ feet.

---

y-axis ***The vertical number line on the coordinate plane.***
Chris looked as if he was rushing down an invisible **y-axis** as he came barreling through the lobby in a vertical line.

Is the *y*-axis vertical or horizontal?

---

y-intercept ***The point where a line intersects the y-axis.***
Chris almost leapt over a group of kids who had planted themselves in his **y-intercept,** but he made it to the door in time.

What is the *y*-intercept of line *n*?

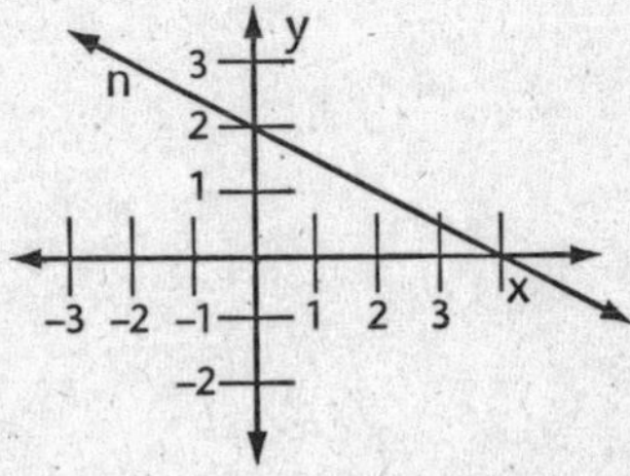

Zero *Neither negative or positive; the additive identity; the smallest digit.*
The whole class cheered when Mr. Stickler informed his class that they'd have **zero** homework for the weekend. Hooray!

Is zero a member of the set of natural numbers?

ANSWERS AND EXPLANATIONS

A

Absolute value

6

Acute angle

(C): 50° is between 0° and 90°, choice (A) is a straight angle, choice (B) is an obtuse angle, and choice (D) is a right angle.

Acute triangle

(C): In an acute triangle, all 3 angles must be acute.

Add (+)

27 + 41 = 68

Adjacent angles

(B) $\angle 1$ and $\angle 2$

Alternate exterior angles

(D): In choice (A), the angles are adjacent and supplementary; in choice (B) the angles are corresponding; and in choice (C) the angles are alternate interior angles.

Alternate interior angles

(C): In choice (A), the angles are adjacent and supplementary; in choice (B) the angles are corresponding; and in choice (D) the angles are alternate exterior angles.

Angle

(A): Because point B is the vertex of this angle, the angle can be named using only letter B or using all 3 letters with B in the middle.

Area

30: The area of a rectangle equals base times height: $10 \times 3 = 30$.

Associative property

(A): Choice (B) is an example of the associative property of addition, choice (C) is an example of the commutative property of multiplication, and choice (D) is an example of the distributive property.

Average

First find the sum of the set of number: $90 + 86 + 81 + 95 = 352$.

Then, divide the sum by the number of values in the set: $352 \div 4 = 88$.

Axes

Points are written in the form (x, y). The parentheses indicate order matters. Because the *x*-coordinate is 4, move 4 units in a positive direction (to the right) on the *x*-axis. Because the *y*-coordinate is 0, do not move up or down. Consequently, the point (4, 0) is on the *x*-axis.

B

Bar graph

(B): Nine students chose baseball as their favorite sport. Four students chose basketball. $9 - 4 = 5$

Base of a polygon

(A): Any side of ΔEFG can be called the base. $\overline{EH}$ is not a side of the triangle. The other 3 line segments are sides of the triangle.

Base of a solid figure

No, a cylinder does not always rest on one of the bases.

Base of an exponent

8: $2^3 = 2 \times 2 \times 2 = 8$

Base Ten

600: Because the 6 is in the hundreds place, its value is $6 \times 100 = 600$.

Binomial

(C): Choices (A) and (B) are monomials, and choice (D) is a trinomial.

Box-and-whisker plot

(C) 24

C

Capacity

Box A: The width and height are too small in box B; the height is too small in box C; and the length, width, and height are all too small in box D.

Celsius

77: $F = \frac{9}{5}C + 32$

$$F = \frac{9}{5}(25) + 32$$

$$F = \frac{9}{5}\left(\frac{25}{1}\right) + 32$$

$$F = \frac{225}{5} + 32$$

$$F = 45 + 32$$

$$F = 77$$

Central angle

(A): In choice (A), the vertex of the angle is at the center of the circle.

Chord

(B): $\overline{GE}$ is a radius, $\overline{CD}$ is a tangent.

Circle

(D): A diameter extends from one side of a circle passing through the center to the other side. Because the circle in choice (D) is the biggest, it has the longest diameter.

Circle graph

(C): Housing is shaded in blue. Because the medium gray section is less than 50% and more than 25%, the best estimate is 35%.

Circumference

The Cascade Carousel: Without even calculating, The Cascade Carousel must have the greatest circumference because it has the longest radius.

Coefficient

(A): Choice B is an exponent, and choice C is a variable.

Common denominator

$\frac{7}{10}$: In order to add two fractions, $\frac{1}{2}$ and $\frac{1}{5}$ must have a common denominator.

To find the common denominator, first list the multiples of each denominator.

Multiples of 2: 2, 4, 6, 8, 10, 12, 14, ...
Multiples of 5: 5, 10, 15, 20, 25, ...

The least common multiple (LCM) is 10.

Write the equivalent fractions with 10 as the denominator.

$$\begin{array}{r} \frac{1}{2} = \frac{5}{10} \\ + \frac{1}{5} = \frac{2}{10} \\ \hline \frac{7}{10} \end{array}$$

Common factor

(D): Factors of 24: **1, 2, 3,** 4, **6,** 8, 12, 24
Factors of 30: **1, 2, 3,** 5, **6,** 10, 15, 30
The common factors of 24 and 30 are **1, 2, 3,** and **6.**

Common multiple

Two children will receive prizes:

Every 5th child receives candy: 5, 10, 15, 20, 25, 30, 35, 40, 45, **50,** 55, 60, 65, 70, 75, 80, 85, 90, 95, **100**

Every 10th child receives a balloon: 10, 20, 30, 40, **50,** 60, 70, 80, 90, **100**

Every 25th child receives a poster: 25, **50,** 75, **100**

The 50th and 100th child receives all three prizes.

Commutative property

The number 3 belongs in the answer blank.

$6 \times 3 = \_\_ \times 6$
$6 \times 3 = 3 \times 6$
$18 = 18$

Compass

(D): All of the above

Complementary angles

$$\begin{array}{rrl} x + 40 = & 90 & \\ -40 = & -40 & \text{(subtract 40 from both sides of} \\ \hline x = & 50 & \text{the equation)} \end{array}$$

Composite number

(B): Factors of 21 are 1, 3, 7, and 21.

Cone

(D)

Congruent

(C): This triangle has the same size and shape as the given triangle.

Consecutive integers

(D): 14, 15, 16

Construction

The process is called a construction.

Coordinates

(2006, 500): Write the horizontal coordinate (x value) first. Then write the vertical coordinate (y value). Use parentheses.

Counting principle

6: If Gabby has 3 shirts and 2 skirts, she has $3 \times 2 = 6$ outfits.

Cross multiplication

$\frac{1}{5} = \frac{4}{x}$

$1 \times x = 5 \times 4$

$x = 20$

Cube

(C)

Cubed

$2^3 = 2 \times 2 \times 2 = 8$

Cube root

$\sqrt[3]{8} = 2$ because $2 \times 2 \times 2 = 8$

Cumulative frequency

Number of Televisions	Tally	Frequency	Cumulative Frequency
1	|||	3	3
2	~~||||~~	5	8
3	~~||||~~ ||	7	15

Customary measurement system

Because 1 yard = 3 feet, 3 yards = 3 × 3 = 9 feet.

Cylinder

(A)

D

Data

Number of Pets	Tally	Frequency
1	|||	3
2	~~||||~~	5
3	~~||||~~ ||	7

Decimal

Look over the list a few times: 1.43, 1.4, and 1.5. 1.4 is equivalent to 1.40. Compare 1.43 and 1.40. Starting at the left, look for the first place where the digits are different.

0 hundredths < 3 hundredths so 1.40 < 1.43

Compare 1.43 and 1.5. Starting at the left, look for the first place where the digits are different.

4 tenths < 5 tenths so 1.43 < 1.5

The order from least to greatest is: 1.4, 1.43, 1.5.

Degree

(B): Choice A is way too small, choice C is a right angle, and choice D is a straight angle.

Denominator

The denominator of this fraction is 20 because there is a total of 20 students in this class.

Diagonal

(C): Choices (A), (B), and (D) are sides of the quadrilateral.

Diameter

(A): Choice (B) is a radius, choice (C) is a chord, and choice (D) is a secant.

Difference

52 – 18 = 34

Digit

4

Discount

The discount is 25% of 40 = 25% × 40 = 0.25 × 40 = \$10; consequently, the shoes cost \$40 – \$10 = \$30.

Distributive property

5(98) = 5(100 – 2) = (5 × 100) – (5 × 2) = 500 – 10 = 490

The distributive property can make a problem easier to compute.

Divide

$$6\overline{)360}^{\,60}$$

Dividend

24: dividend ÷ divisor = quotient

Divisible

No, because when 42 is divided by 5, there is a remainder.

$$8\overline{)42}^{\,5}\ \text{R2}$$

Divisor

4: dividend ÷ divisor = quotient

E

Element

There are 5 elements in the set.

Endpoint

This segment can be called $\overline{AB}$ or $\overline{BA}$.

Equals

4: Find the value of x in the equation $x + 3 = 7$ by subtracting 3 from both sides.

$x + 3 - 3 = 7 - 3$

$x = 4$

Check. $4 + 3 = 7$

Equation

(D): Choice (A) is an expression, choices (B) and (C) are inequalities.

Equilateral triangles

Yes: All equilateral triangles are similar because they all contain three 60° angles. Consequently, equilateral triangles have the same shape.

Equivalent fractions

There are an infinite number of answers to this question. Each answer can be found by multiplying the numerator and denominator by the same number. Here is a list of some of the possible answers:

$$\frac{2}{4}, \frac{3}{6}, \frac{4}{8}, \frac{5}{10}, \frac{6}{12}, \frac{7}{14}, \frac{8}{16}, \frac{9}{18}, \text{ and } \frac{10}{20}.$$

Estimate

First, add the values of the front digits:

$$\begin{array}{rcr} 3.72 & \rightarrow & 3 \\ 2.19 & \rightarrow & 2 \\ 9.06 & \rightarrow & 9 \\ \underline{4.97} & \rightarrow & \underline{+\ 4} \\ & & 18. \end{array}$$

Next, estimate the sum of the remaining digits:

$0.72 + 0.19 + 0.06 + 0.97 =$ about 2.

Lastly, combine your results: $18 + 2 = 20$.

Even number

(C): 47 is an odd number, not an even number.

Exponents

(B): $100{,}000 = 10 \times 10 \times 10 \times 10 \times 10 = 10^5$

Expression

(A): Choice (B) and (C) are an inequalities, and choice D is an equation.

Exterior angle of a polygon

(D): All of the other choices are interior angles.

F

Faces

The figure has 6 faces.

Factor

All the factors of 20 are 1, 2, 4, 5, 10, and 20.

Factorial

$4! = 4 \times 3 \times 2 \times 1 = 24$

Factoring

$32 = 1 \times 32$
$32 = 2 \times 16$
$32 = 4 \times 8$

Fahrenheit

$$F = \frac{9}{5}C + 32$$

$$F = \frac{9}{5}10 + 32$$

$$F = \frac{9}{5}\left(\frac{10}{1}\right) + 32$$

$$F = \frac{90}{5} + 32$$

$$F = 18 + 32$$

$$F = 50$$

FOIL

$101 \times 101 = (100 + 1)(100 + 1) =$

$100 \overset{F}{\times} 100 + 100 \overset{O}{\times} 1 + 1 \overset{I}{\times} 100 + 1 \overset{L}{\times} 1$

$10{,}000 + 100 + 100 + 1 = 10{,}201$

Formula

$A = \frac{1}{2}bh$

Fraction

$\frac{2}{6}$: Out of six shapes, 2 have 5 sides.

Frequency

$\frac{7}{15}$

G

Gallon

Because 1 gallon equals 8 pints, 5 gallons equals 5 × 8 pints = 40 pints.

Gram

Because 1 kilogram equals 1,000 grams, 1 kilogram is larger than 1 gram.

Greater than

(B): 5 > 3

Greater than or equal to

(D): 2 × 3 ≥ 1 × 5

6 ≥ 5 (6 is greater than or equal to 5 is a true statement.)

Greatest common factor

Factors of 28: 1, 2, 4, 7, 14, 28

Factors of 42: 1, 2, 3, 6, 7, 14, 21, 42

While 2 and 7 are both common factors, 14 is the greatest common factor.

H

Height (altitude)

(C): $\overline{BE}$ is the altitude because it is perpendicular to the base.

Hexagon

(C): A hexagon has 6 sides.

Histogram

(B): 4

Horizontal

The *x*-axis is horizontal.

Hundreds

(A): 3

Hundredths

(D): 8

Hypotenuse

Side $\overline{AB}$, also called *c*, is the hypotenuse. Sides $\overline{BC}$ and $\overline{AC}$ also called *a* and *b* respectively, are the legs.

I

Identity

Answers will vary. Here are three possible answers:

$1 + 0 = 1$
$2 + 0 = 2$
$3 + 0 = 3$

Improper fraction

(B): $\frac{7}{4}$

Increase

$4 + 3$

Inequality

$2 + 4 \times 3 < \frac{20}{2} + 3 \times 5$

Use order of operations: PEMDAS

$2 + 12 < 10 + 15$

$14 < 25$

Infinity

No, infinity is larger than any number you can imagine.

Integer

(B): Choice (A) simplifies to 4, which is an integer; choice (C) is an integer; and choice (D) equals 8, which is an integer.

Inverse

The additive inverse of 4 is –4 because $4 + -4 = 0$.

Irrational

(D): $\sqrt{7} \approx 2.6457513...$; this decimal does not terminate or repeat.

Isolate

Isolate x by adding 2 to both sides of the equation:

$$\begin{aligned} x - 2 &= 10 \\ +2 &= +2 \\ x &= 12 \end{aligned}$$

Isosceles trapezoid

(C): While all choices are trapezoids, only choice (C) is an isosceles trapezoid; the marks on the legs indicate they are congruent.

Isosceles triangle

(B): It's the only triangle with 2 equal sides.

L

Least common denominator

$\frac{2}{3}$: In order to add or subtract fractions, find the least common multiple of the 2 denominators. The least common multiple of 2 and 6 is 6.

Multiples of 2: 2, 4, **6**, 8, 10, 12, ...
Multiples of 6: **6**, 12, 18, 24, 30, ...

$$\frac{1}{2} = \frac{1 \times 3}{2 \times 3} = \frac{3}{6}$$

$$\frac{3}{6} + \frac{1}{6} = \frac{4}{6}$$

$$\frac{4}{6} = \frac{2}{3}$$

Least common multiple

Multiples of 5: 5, 10, 15, 20, 25, **30**, 35, 40, . . .
Multiples of 6: 6, 12, 18, 24, **30**, 36, 42, 48, . . .
Multiples of 10: 10, 20, **30**, 40, 50, 60, 70, . . .

The smallest number that is a multiple of 5, 6, and 10 is 30.

Leg

Sides $\overline{BC}$ and $\overline{AC}$, also called *a* and *b* respectively, are the legs; Side $\overline{AB}$, also called *c*, is the hypotenuse.

Length

Because the length equals 6 and the width equals 2, the length is $6 - 2 = 4$ cm longer than the width.

Less than

$38 < 271$

Less than or equal to

$509 < 567 < 576 \leq 576 < 5{,}670$

Like terms

(D): The terms in choice (D) contain the same variable and exponent.

Line

$\overleftrightarrow{AB}$, $\overleftrightarrow{AC}$, $\overleftrightarrow{BC}$, $\overleftrightarrow{BA}$, $\overleftrightarrow{CA}$, $\overleftrightarrow{CB}$, or *j*.

Line graph

Approximately 60 people; look at the graph halfway between June 14 and June 21 and find the corresponding number of swimmers.

Line segment

$\overline{DE}$, $\overline{DF}$, $\overline{EF}$, $\overline{FD}$, $\overline{ED}$, or $\overline{FE}$

Line symmetry

(D): Choices (A) and (C) have a vertical line of symmetry; choice (B) has a horizontal line of symmetry.

Linear equations

(0, 1): Find the point of intersection; write the *x*-coordinate first and the *y*-coordinate second; use parentheses.

Liter

Because a milliliter equals of $\frac{1}{1,000}$ a liter, multiply the number of liters by 1,000 to determine the number of milliliters.

$$0.01 \text{ liters} = 0.01 \times 1,000 = 10 \text{ milliliters}$$

Lowest terms

$$\frac{5}{15} = \frac{5 \times 1}{5 \times 3} = \frac{\cancel{5} \times 1}{\cancel{5} \times 3} = \frac{1}{3}$$

M

Mean

Add the numbers and divide by 5.

$$10 + 8 + 5 + 9 + 18 = 50$$
$$50 \div 5 = 10$$

Median

(A): Mean > mode.

First place in order: 6, 6, 7, 8, 11.

Mode = 6 (it occurs the most)

Median = 7 (middle number when placed in order)

$$\text{Mean} = \frac{6 + 6 + 7 + 8 + 11}{5} = \frac{38}{5} = 7.6$$

Meter

First convert centimeters to meters by dividing by 100. Then convert meters to kilometers by dividing by 1,000.

$$50{,}000 \text{ cm} \div 100 = 500 \text{ m}$$
$$500 \text{ m} \div 1{,}000 = 0.5 \text{ km}$$

Number sense will help determine if an answer is reasonable and will help in remembering whether to multiply or divide. When converting to a larger unit, the answer gets smaller. If a distance is given in centimeters, it takes fewer kilometers to equal that distance.

Metric system

Compare each measurement to 1 meter.

1 millimeter < 1 centimeter < 1 meter < 1 kilometer

Midpoint

$\overline{GH} \cong \overline{HI}$

Mile

Because 1 mile = 5,280 feet and 3 feet = 1 yard, divide the number of feet by 3 to get the number of yards.

$$5{,}280 \div 3 = 3\overline{)5{,}280}^{\;1{,}760}$$

1 mile = 1,760 yards

Minus

63 minus 41 = 63 − 41 = 22

Mixed number

$\frac{9}{4}$ means 9 ÷ 4, $4\overline{)9}$ → 2 R 1

Write the quotient as the whole number.

Write the remainder in the numerator of the fraction. The divisor is the denominator. 2 R 1 is written as $2\frac{1}{4}$.

Mode

Because it occurs the most often, 91 is the mode.

Monomial

(A): Choices (B) and (D) are binomials, and choice (C) is a trinomial.

Multiple

The first 5 multiples of 3 are 3, 6, 9, 12, and 15.

Multiply

$$\begin{array}{r} 15 \\ \times\ 3 \\ \hline 45 \end{array} \quad \text{OR} \quad 15 + 15 + 15 = 45$$

N

Natural numbers

(D): 1,245

Negative numbers

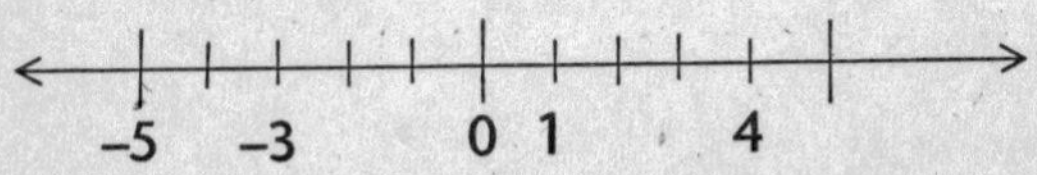

Number line

6: Because 6,438,689 is closer to 6,000,000 than 7,000,000 on the number line, it rounds to 6,000,000.

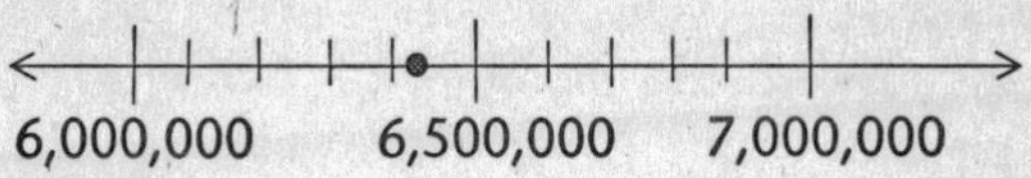

Numerator

The numerator of this fraction is 7 because there are a total of 7 students with brown eyes.

Obtuse angle

(D): Choice (A) is an acute angle, choice (B) is a right angle, and choice (C) is a straight angle.

Obtuse triangle

(B): Choice (A) is a right triangle, and choices (C) and (D) are acute triangles.

Octagon

(D): Choice (A) is a triangle, choice (B) is a quadrilateral, and choice (C) is a hexagon because it has 6 sides.

Odd number

(A): 41 is odd because it ends in a 1.

Ones place

In 5,692.48, the number 2 is in the one's place.

5,692.48
5 is in the thousands place
6 is in the hundreds place
9 is in the tens place
2 is in the ones place
4 is in the tenths place
8 is in the hundredths place

Operations

Addition: 6 + 10 = 16

Order of operations

12 + 5 × 2 (multiply first)
12 + 10 (then add)
22

Outcome

2: Heads or Tails

P

Parallel

j is parallel to k ($j \parallel k$).

Parallelogram

(A): Choices (B) and (D) are trapezoids, and choice (C) is a triangle.

Parentheses

Simplify inside parentheses first, then multiply:

$$5 \times (3 + 1) = 5 \times 4$$
$$= 20$$

Pattern

The missing number is 32; each number is doubled.

Pentagon

(C): Choice (A) is a triangle, choice (B) is a quadrilateral, and choice (D) is a hexagon.

Percent

0.6 (read as 6 tenths) $= \frac{6}{10} = \frac{6}{100} = 60\%$

Percent of decrease

Decrease: 800 – 600 = 200

Original: 800

Ratio: $\frac{200}{800}$ = .25 = 25%

Percent of increase

Increase: 700 – 500 = 200

Original: 500

Ratio: $\frac{200}{500}$ = 0.4 = 40%

Perfect squares

1, 4, 9, 16, 25, 36, 49, 64, 81, and 100

(The number 1 is a perfect square because 1^2 = 1; 4 is a perfect square because $2^2 = 4$; 9 is a perfect square because $3^2 = 9$; and so on.)

Perimeter

P = length + width + length + width

P = 5 + 3 + 5 + 3 = 16

Perpendicular

$\overleftrightarrow{CD}$, is perpendicular to $\overleftrightarrow{EF}$, ($\overleftrightarrow{CD} \perp \overleftrightarrow{EF}$).

Pi

Because pi is equal to the ratio of the circumference to the diameter, $\pi = \frac{C}{d}$.

Cross multiply to get $C = \pi d$.
Because $d = 5$, $C = \pi \times 5 = 5\pi = 5 \times 3.14 = 15.7$.

Pint

4 pints × 2 cups = 8 cups

Place value

50 because 5 is in the tens place (5 × 10 = 50).

Plane

A plane is named using three noncollinear points; there are 6 possible answers; plane ABC, plane ACB, plane BAC, plane BCA, plane CAB, or plane CBA.

Plot

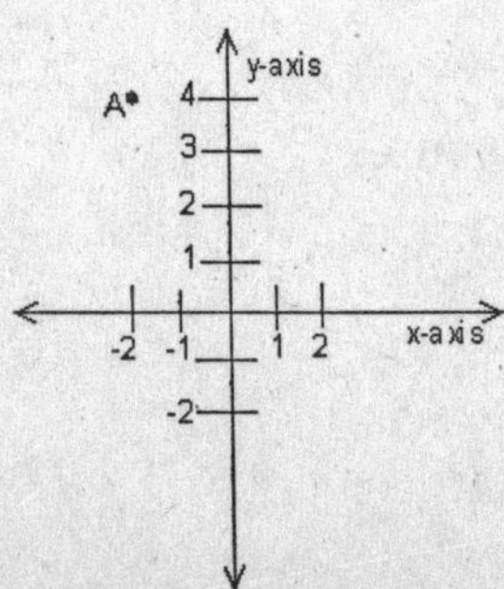

Point

Point C is located on circle A; point A is at the center of the circle; point B is located in the interior of the circle; point D is located in the exterior of the circle.

Polygon

(A): Choices (B) and (D) are not closed figures; and choice (C) is a circle; circles are not polygons because they have arcs, not line segments.

Polynomial

Yes: a polynomial can have any amount of terms.

Positive

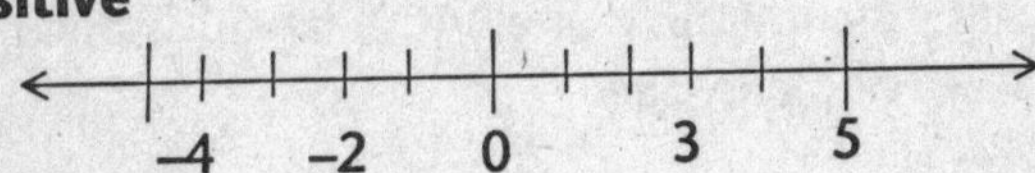

Pound

1 pound = 16 ounces

1 cup = 8 ounces

Therefore, 1 pound = 2 cups.

Power

Nothing happens because any number raised to the first power equals itself. For example, $7^1 = 7$.

Prime number

(A): 5 is prime because it has only 2 factors, 1 and itself; choices (B), (C), and (D) are all composite because they have 3 or more factors.

Factors of 5: 1 and 5
Factors of 9: 1, 3, and 9
Factors of 20: 1, 2, 4, 5, 10, and 20
Factors of 21: 1, 3, 7, and 21

Prime factorization

$100 = 5 \times 5 \times 2 \times 2$

```
   100
   | |
 25   4
 | |  | |
 5 5  2 2
```

Prism

(D): Cones and pyramids are not prisms because they come to a point; prisms must have 2 congruent, parallel bases.

Probability

Number of desired outcomes (getting a 5) = 1

Total number of possible outcomes = 6

$$P(E) = \frac{\textit{number of desired outcomes}}{\textit{total number of possible outcomes}} = \frac{1}{6}$$

Product

Product means multiply.

$4 \times 7 = 28$

Proper fraction

(C): Choices (A), (B), and (D) are improper fractions because the numerator is bigger than the denominator.

Proportion

$\frac{2}{5} = \frac{2 \times n}{20}$ Because 20 = 5 × 4, the denominator has been multiplied by 4. Consequently the numerator must be multiplied by 4.

$$n = 4$$

Protractor

45°: The measure of this angle can be found using a protractor or by estimating; it appears to be about half of a right angle.

Pyramid

The total number of edges is 8.

Pythagorean Theorem

No, the Pythagorean Theorem only works in right triangles.

Q

Quadrants

To find the location of this point, start at the origin (0, 0). Move 5 units to the right and 1 unit down. Point A is in quadrant IV.

Quadrilateral

(B): Choice (A) is a triangle, and choice (C) is a pentagon, and choice (D) is a hexagon.

Quart

2 quarts = 8 cups

Because 1 quart equals 4 cups, 2 quarts equals $4 \times 2 = 8$ cups.

Quotient

The quotient is 6, or the result of division.

R

Radical

$\sqrt{16}$ equals 4 because $4 \times 4 = 16$.

Radicand

(B): 3 appears under the radical sign.

Radius

(B): Choice A is a diameter, choice C is a chord, and choice D is a secant.

Raised to

This expression translates to 2^3, which is equal to $2 \times 2 \times 2 = 8$.

Range

Highest test score: 97
Lowest test score: 80
Range = 97 – 80 = 17

Rate

rate = 30 miles per hour = 30 m/h
time = 2 hours
distance = rate × time
distance = 30 × 2 = 60 miles

Ratio

(C): The ratio 2:3 means for every 2 circles, there are 3 squares. Consequently, for every 4 (2 × 2) circles, there are 6 (3 × 2) squares.

Rational

(C): Choices (A), (B), and (D) are irrational because they do not terminate or repeat.

Ray

(B): Because the endpoint of the ray must be listed first, choice (B) cannot be used to name the ray.

Real numbers

$\sqrt{9} = 3$

$\sqrt{2}$ is between 1 and 2 because $\sqrt{1} = 1$ and

$\sqrt{4} = 2$

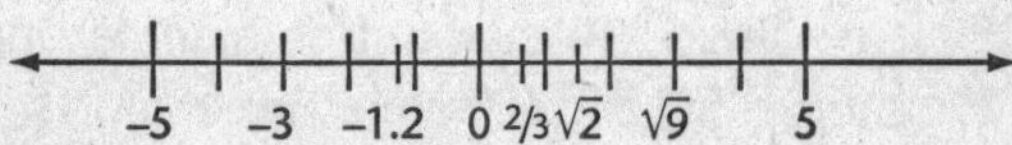

Rectangle

(D): In a rectangle, opposite sides must be equal, but all 4 sides are not always equal.

Reflection

(C): When $\overline{AB}$ is reflected over line *k*, the corresponding vertices of the two segments must be on opposite sides of line *k* and the same distance from *k*.

Remainder

The remainder is 2 because $5\overline{)17}$ with quotient 3, Remainder 2.

Repeating decimal

$72.151515151515... = 72.\overline{15}$

Rhombus

(C): A rhombus must have 4 *equal* sides.

Right angle

(A): Choice (B) is an acute angle and choices (C) and (D) are obtuse angles.

Right triangle

(A): Choice (B) is an obtuse triangle, and choices (C) and (D) are acute triangles.

Rotation

(C): $\triangle ABC$ is turned about the origin.

Rotational symmetry

(D): S has rotational symmetry because it looks identical to itself after a rotation of 180°, choice (A) has a vertical line of symmetry, and choice (C) has a horizontal line of symmetry.

Round

Look at the digit one unit to the right of the tenths place (the hundredths place). If that digit is greater than or equal to 5, round up. Because this digit is 2 (not greater than or equal to 5), round down. 46.82 rounded to the nearest tenth equals 46.8.

S

Sales tax

To compute the tax, take 8% of 20 (of means multiply).

$$\begin{array}{rcrl} 20 & \rightarrow & 20 & \\ \times\ 8\% & \rightarrow & \underline{\times\ 0.08} & \text{convert \% to decimal} \\ & & \$1.60 & \text{sales tax} \end{array}$$

Original cost of the shirt plus sales tax = final cost.

$$\begin{array}{rcrl} 20 & \rightarrow & 20.00 & \\ \underline{+\ 1.60} & \rightarrow & \underline{+\ 1.60} & \text{line up the decimals} \\ \text{total cost:} & & \$21.60 & \text{when adding or} \\ & & & \text{subtracting} \end{array}$$

Scalene

(A): It's an equilateral triangle.

Scientific notation

Write as the product of 2 numbers. The first number is between 1 and 10 and the second is a power of 10.

$$491{,}000{,}000 = 4.91 \times 108$$

Semicircle

(B): Because $\overset{\frown}{BDC}$ is half of the circle, it is a semicircle.

Set

(B): Choice (A) represents even numbers, choice (C) represents numbers divisible by 3, and choice (D) represents numbers divisible by 4.

Similar

(D): These triangles have the same shape, just different sizes.

Simple interest

$I = p \times r \times t$

$I = 2{,}000 \times 0.07 \times 5 = \$700$

Note: Don't forget to convert 7% to 0.07.

Simplify

Use order of operations (PEMDAS).

$5^2 - 4 \times 6$ (Exponents first)

$25 - 4 \times 6$ (Multiplication next)

$25 - 24$ (Subtract)

$= 1$

Slope

Use *slope* = $\frac{rise}{run}$ as a way to help in remembering that slope is the ratio of the change in *y*-coordinates (rise) to the change in *x*-coordinates (run). If the rise is positive, go up. If the rise is negative, go down. If the run is positive, move to the right. If the run is negative, move to the left.

Pick a point on the line. Because slope = $\frac{2}{2}$ = $\frac{rise}{run}$, move up 2 and to the right 3 to find the next point on the line. Only choice (C) has a rise of 2 and run of 3.

Solid

(A): A cylinder has 2 circular bases; choices (B), (C), and (D) are prisms.

Sphere

(B): Choice (A) is a cube, choice (C) is a cylinder, and choice (D) is a cone.

Square

If one side of a square equals 5, all 4 sides equal 5. $P = s + s + s + s = 4s = 4 \times 5 = 20$.

Squared

The expression "7 squared" means $7^2 = 7 \times 7 = 49$.

Stem-and-leaf plot

7 | 9 means 79

Stem	Leaf
7	8 9 9
8	1 5 5
9	3 3 3 8

Straight angle

(C): Choice (A) is an acute angle, choice (B) is a right angle, and choice (D) is an obtuse angle.

Subtract

Subtract 10 from 25 means take 10 away from 25.

$25 - 10 = 15$

Substitute

To check if $x = 6$ in the equation $x + 4 = 10$, substitute 6 in for x.

$x + 4 = 10$

$6 + 4 = 10$

$10 = 10$

$x = 6$ (Because this is a true statement, is the correct solution.)

Sum

Sum means add.

$25 + 15 = 40$

Supplementary angles

$x + 80 = 180$

$-80 \quad -80$ (subtract 80 from both sides of the equation)

$x = 100$

Surface area

A cube has 6 equal square shaped faces. To find the area of one face, use the formula for the area of a square, $A = s^2$.

$$A = s^2 = 3^2 = 3 \times 3 = 9$$

Because there are 6 faces, multiply 9 by 6 to find the surface area.

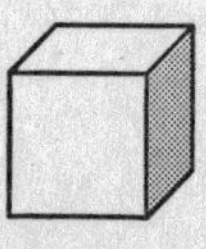

$9 \times 6 = 54$

T

Tens

In 598,491, the number 9 is in the tens place.

Tenths

321.5

Look at the digit one unit to the right of the tenths place (the hundredths place). If that digit is greater than or equal to 5, round up. Since this digit is 8, round up. 321.48 rounded to the nearest tenth equals 321.5.

Terminating decimal

(A): Choices (B) and (C) are repeating decimals, and choice (D) is a nonterminating decimal.

Term

There are 4 terms in the expression.

Thousandths

In 9,346.2875, 7 is in the thousands place.

9,346.2875
9 is in the thousands place
3 is in the hundreds place
4 is in the tens place
6 is in the ones place
2 is in the tenths place
8 is in the hundredths place
7 is in the thousandths place

Three-dimensional

(C): Choices (A), (B), and (D) are two-dimensional.

Times

Times means multiply.
3 × 7 = 21

Ton

3 tons = 6,000 pounds
Because 1 ton = 2,000 pounds, 3 tons = 3 × 2,000 = 6,000 pounds.

Transformation

(D): Choice (A) changes the size of a figure, choice (B) "flips" a figure over a line, and choice (C) "turns" a figure about a point.

Translation

(D): Choice (A) is a dilation, choice (B) is a reflection, and choice (C) is a rotation.

Transversal

Line *m* is a transversal.

Trapezoid

(C): Choice (A) is a parallelogram, choice (B) is a rectangle, and choice (D) is a square.

Tree diagram

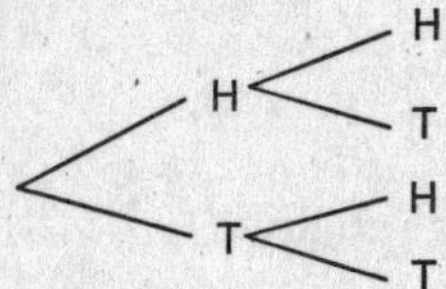

Triangle

To classify a triangle by its angles, determine if the triangle is acute, right, or obtuse. To classify a triangle by its sides, determine if the triangle is equilateral, isosceles, or scalene. This triangle is acute because all of its angles are less than 90° and isosceles because 2 of its sides are equal.

Trinomial

(D): Choices (A) and (B) are monomials, and choice (C) is a binomial.

Two-dimensional

(A): Choices (B), (C), and (D) are three-dimensional figures.

Variable

(D): *y* is the variable, –8 is the coefficient, and 2 is the exponent.

Venn Diagram

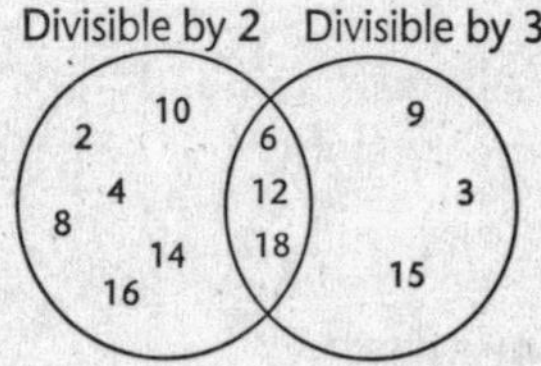

Vertex

Point B is the vertex of $\angle ABC$ because it is the common endpoint of $\overrightarrow{BA}$ and $\overrightarrow{BC}$.

Vertical

The vertical axis in this graph is the one that shows the number of swimmers.

Vertical angles

$\angle 1$ and $\angle 3$ are vertical angles.
$\angle 2$ and $\angle 4$ are also vertical angles.

Volume

(C): The measure of the third dimension, height, must be given in order to find the volume.

W

Weight

(B): Gallons may be used when measuring volume, not weight.

Whole numbers

(A): The set of whole numbers does not include negative numbers.

Width

Area = length × width

$A = lw$

Length = 10

Width = 6

$A = 10 \times 6 = 60$

x-axis

The *x*-axis is horizontal.

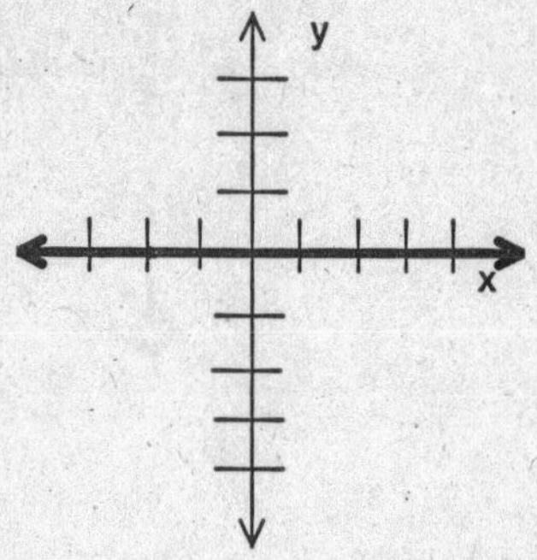

x-intercept

The *x*-intercept is 1. The line crosses the *x*-axis at (1, 0).

Y

Yard

4 yards = 12 feet

If 1 yard = 3 feet, 4 yards = $4 \times 3 = 12$ feet.

***y*-axis**

The *y*-axis is vertical.

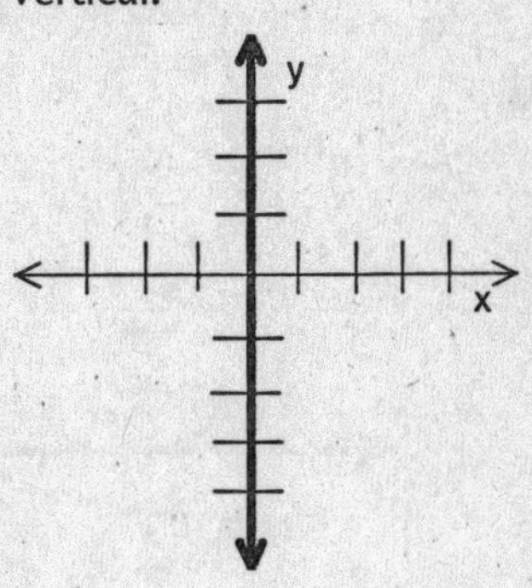

***y*-intercept**

The y-intercept is 2. The line crosses the *y*-axis at (0, 2).

Z

Zero

Zero is not a natural number.
Natural numbers include {1, 2, 3, 4,...}.